NANOTECHNOLOGY SCIENCE AND TECHNOLOGY

SURFACE PLASMON RESONANCE

Nanotechnology Science and Technology

Additional books in this series can be found on Nova's website under the Series tab.

Additional e-books in this series can be found on Nova's website under the e-book tab.

NANOTECHNOLOGY SCIENCE AND TECHNOLOGY

SURFACE PLASMON RESONANCE

PRANVEER SINGH

New York

For permission to use material from this book please contact us:
Telephone 631-231-7269; Fax 631-231-8175
Web Site: http://www.novapublishers.com

Library of Congress Cataloging-in-Publication Data

ISBN: 978-1-63321-835-2

Published by Nova Science Publishers, Inc. † New York

CONTENTS

PREFACE

SPR is real-time, label-free measurements of binding kinetics and affinity. This has distinct advantage over radioactive or fluorescent labeling methods, in terms of 1) ligand-analyte binding kinetics, that can be probed without costly and time-consuming labeling process that may interfere with molecular binding interactions; 2) binding rates and affinity can be measured directly and 3) low affinity interactions in high protein concentrations can be characterized with less reagent consumption than other equilibrium measurement techniques; 4) Label-free detection of molecular interactions presents an attractive alternative. This is particularly true for protein targets where labels can compromise protein function. Label-free methods are also desirable because these are typically compatible with real-time detection, enabling the determination of rates of association. The success of SPR biosensor was indicated by the growing number of commercially available instruments. Since Biacore AB (originally Pharmacia Biosensor AB) launched the first commercial SPR biosensor in 1990, there have been many more competing SPR instruments including IASys (Affinity Sensors), SPR-670 (Nippon Laser Electronics), IBIS (IBIS Technology BV), TISPR (Texas Instruments), etc.

Plasmon resonance techniques such as SPR, SPR-imaging, SPR-MS (mass spectrometry) and PWR, trends in protein array technology and a potential use of SPR biosensors in proteome and optics based research will be reviewed in terms of their fundamentals, and in the latest applications with emphasis on studies performed with membrane proteins. This work will also throw light on SPR sensors (from typical Kretschmann prism configurations to fiber sensor schemes) with micro- or nano-structures for local light field enhancement, extraordinary optical transmission, interference of surface plasmon waves, plasmonic cavities, etc. Additionally, a note on covalent and non-covalent immobilization methods for the attachment of the molecules to the sensor surface is also reviewed for the functional reconstitution of membrane proteins in various types of lipid mono/bilayers. Advantages and disadvantages of each methodology are provided along with some of the latest accomplishments with emphasis on the area of GPCRs.

With ever increasing advances in high-throughput proteomics and system biology, there is a growing demand for the instruments that can precisely quantitate a wide range of receptor-ligand binding kinetics in a high-throughput fashion. The ability to quantify kinetics and affinities of receptor-ligand binding interactions is essential for basic biology, viz. ligand screening, immunology, cell biology, signal transduction, and nucleotide-nucleotide, nucleotide-protein, protein-protein, protein-lipid interactions, biomarker discovery, proteomics, pharmaceutical development and drug discovery, among others. Optical biosensors that utilize surface plasmon waves (SPs) – electromagnetic waves propagating on the interface between a metallic film and a dielectric medium – have been widely used for this purpose. Techniques based on surface plasmons such as Surface Plasmon Resonance (SPR), SPR Imaging, Plasmon Waveguide Resonance (PWR) and others, have been increasingly used to determine the affinity and kinetics of a wide variety of real time molecular interactions such as protein-protein, lipid-protein and ligand-protein, without the need for a molecular tag or label.

Dr. Pranveer Singh
Assistant Professor
Department of Zoology
Indira Gandhi National Tribal University (IGNTU)
Amarkantak - 484886 (MP)
Tel: +91-7629-269771
Tel: +91-9424930522
E-mail: mailpranveer@gmail.com

Chapter 1

INTRODUCTION

OVERVIEW OF SURFACE PLASMON RESONANCE (SPR)

Surface Plasmon Resonance (henceforth referred to as SPR) is an optical phenomenon that provides a non-invasive, label-free means of observing binding interactions between an injected analyte and an immobilized biomolecule (ligand) in real time. The SPR effect is sensitive to binding of analyte because the associated increase in mass causes a proportional increase in refractive index, which is observed as a shift in the resonance angle. A flow injection analysis configuration is commonly employed in which the analyte of interest, solvated in a buffer solution, is transported across the sensing surface, where it interacts with the immobilized biomolecule.

SPR can be described as the resonant, collective oscillation of valence electrons in a solid stimulated by incident light. The resonance condition is established when the frequency of light photons matches the natural frequency of surface electrons oscillating against the restoring force of positive nuclei. When planar surface is involved, then excitation of surface plasmon (Surface plasmons, also known as polaritons, are surface electromagnetic waves that propagate in a direction parallel to the metal/dielectric (or metal/vacuum) interface) by light is called SPR, while if nanometer sized metallic surface is involved then it is Localised surface plasmon resonance (LSPR). The existence and properties of surface plasmons can be described by quantum theory or Drude model [1, 2].

SPR is the basis of many standard tools for measuring adsorption of material onto planar metal (typically gold and silver) surfaces or onto the surface of metal nanoparticles. It is the fundamental principle behind many color-based biosensor applications and different lab-on-a-chip sensors.

When combined with appropriate surface chemistry, micro fluidics and software, this technique is unmatched in its range of applications including:

- Affinity analysis
- Kinetic analysis
- Concentration assays
- Active concentration assays
- Binding stoichiometry
- Thermodynamic analysis
- Study of interaction mechanisms
- Dependence of interaction on environmental conditions
- Routine screening
- Ligand-fishing
- Epitope mapping

1.1. Physical Basis of Surface Plasmon Resonance

SPR transducers are usually constructed by using prism coupling of incident light onto an optical substrate that is coated with a semi-transparent noble metal under conditions of total internal reflection. First, a condition of total internal reflection (Figure 1) must exist at the interface. Total internal reflection will exist until the incident angle decreases to θc (critical angle), at this angle; some of the light is refracted across the interface. At the point of reflection at the interface, an evanescent field (standing wave) will penetrate the exit medium to a depth in the order of 1/4 of the incident light wavelength. [3]

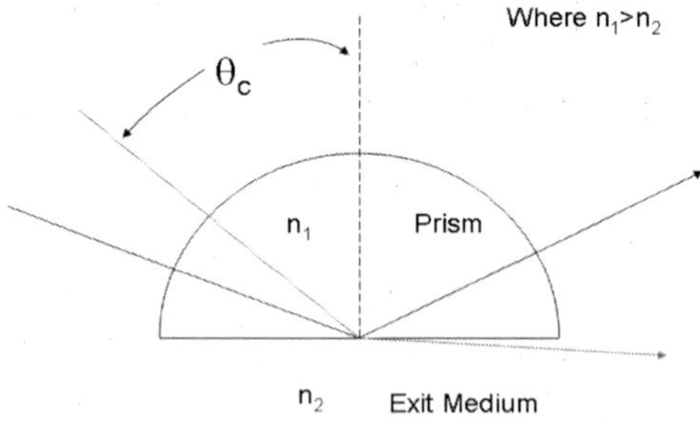

Figure 1. Total internal reflection at a prism interface.
(http://www.sensiqtech.com/uploads/file/support/spr/Overview_of_SPR.pdf)

Here, n_1 and n_2 are the refractive indexes of the prism and exit medium, respectively. The refracted beam at the critical angle is shown in red. All light is reflected at incident angle greater than θc.

If a semi-transparent noble metal film is placed at the interface (Figure 2), then under conditions of total internal reflection surface plasmon resonance can occur. This is commonly known as the *Kretschmann configuration* [4]. There are various methods that use different coupling devices between the incident light and surface Plasmon. The mostly widely used method is the Kretschmann geometry where the metallic film is coated directly on the prism or on a plate held on a prism with an appropriate index matching liquid.

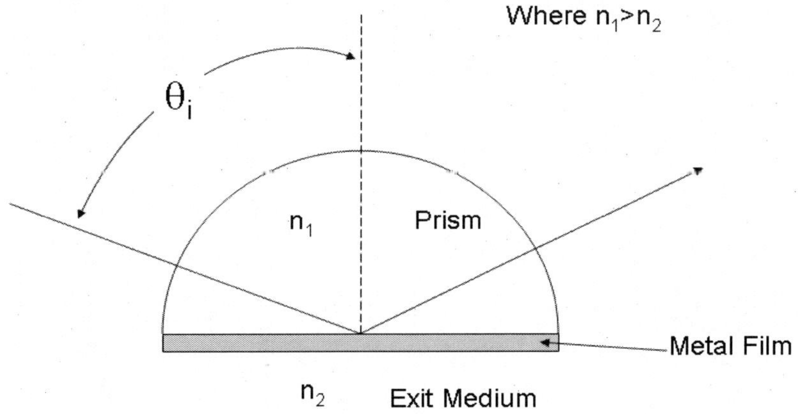

Figure 2. SPR using a prism coupling configuration. (http://www.sensiqtech.com/uploads/file/support/spr/Overview_of_SPR.pdf)

A condition of surface plasmon resonance will occur when the following conditions are satisfied. The incident wave vector is given by the following expression:

$$K_i = (2\pi/\lambda)\, n \sin\theta_i$$

where K_i is a component of the incident light wave vector parallel to the prism interface, θi is the incident light angle, λ is the wavelength of the incident light and n is the refractive index of the prism [5,6].

The wave vector of the plasmon mode is described by the following expression

$$K_p = (2\pi/\lambda)(\varepsilon_1 \varepsilon_2/\varepsilon_1 + \varepsilon_2)^{1/2}$$

where Kp is the surface plasmon wave vector and ε_1 and ε_2 are the dielectric permittivity constants of the metal film and the dielectric exit medium, respectively [7].

Typical metals that support surface plasmons are silver and gold, but metals such as copper, titanium or chromium have also been used.

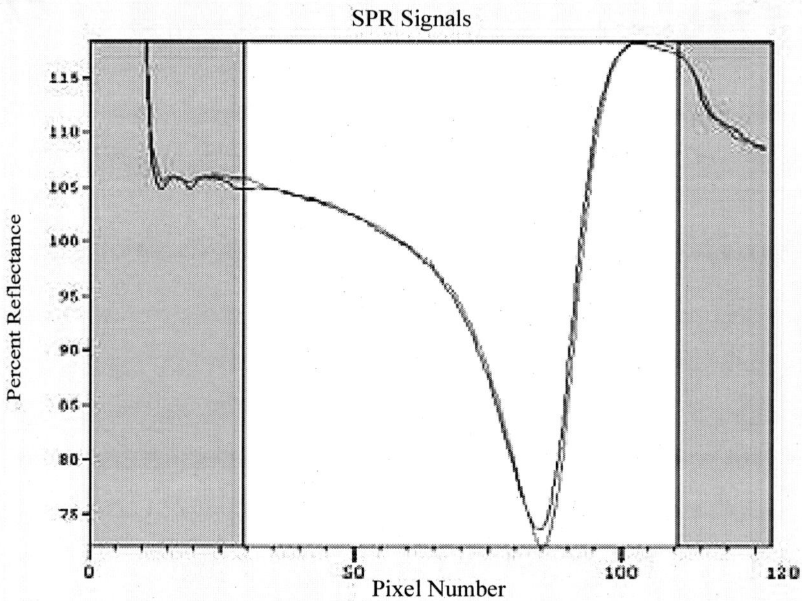

Figure 3. SensíQ Dual Channel SPR Minima.
(http://www.sensiqtech.com/uploads/file/support/spr/Overview_of_SPR.pdf)

Surface plasmon resonance occurs when the intensity of the reflected light will decrease where surface plasmon resonance exists thereby giving rise to a well defined minimum in the reflectance intensity. If the incident angle is fixed and polychromatic light is reflected from the surface, then light will be absorbed by the resonance at particular wavelengths giving rise to a typical plasmon resonance minimum in the reflectance spectrum. If monochromatic light is reflected from the surface over a range of incident angles then a similar reflectance minimum will occur with respect to the incident angle. The reflectance minimum that results from plasmon resonance is caused by the phase difference of the surface mode relative to the incident photon field. In fact the phase difference below the resonance is 0° and approaches 180° above the resonance. Therefore, the photons reflecting from the metal-prism interface undergo destructive interference with the photons emitted by the excited plasmons (being 180° out of phase just above the maximum resonance) and this result in the characteristic reflectance minimum. Figure 3 shows a typical SensíQ SPR minimum.

1.2. Step by Step Working of SPR

- Polarized light is shined through a prism onto a metal (Ag, Cu etc.) film (Figure 4a,b)
- Light is rotated along prism past the critical angle until total internal reflection is observed (Figure 4a, b)
- Maximum loss of reflected light intensity at the Resonance Angle (Figure 4a, b)
- At resonance angle photons from the light source reach beyond surface of prism, causing excitation and absorption parallel to the metal layer (Figure 4a, b)
- These are called electron density waves or Surface Plasmon (Figure 4a, b)
- When photons are converted into surface plasmon, reflected light intensity decreases (Figure 4a, b)
- As analyte binds to the ligand it alters the refractive properties behind the metal film, altering the resonance angle (Figure 4a, b)
- The difference in the refractive angle is plotted as a function over time in a Sensorgram (Figure 4c)

For biosensing, it is this change in the refractive index of the dielectric exit medium at the gold surface that is of interest. Therefore, if the refractive index of the prism is constant then a change in the resonance condition may be related to the changes in the refractive index of the exit medium. In this way, it is possible to monitor the accumulation of biofilms on the gold surface in order to measure binding of biomolecules to a surface that has been coated with an affinity ligand. This SPR configuration may be considered a surface sensitive refractometer where the sensitivity depth is defined by the penetration depth of the evanescent field (~200nm).

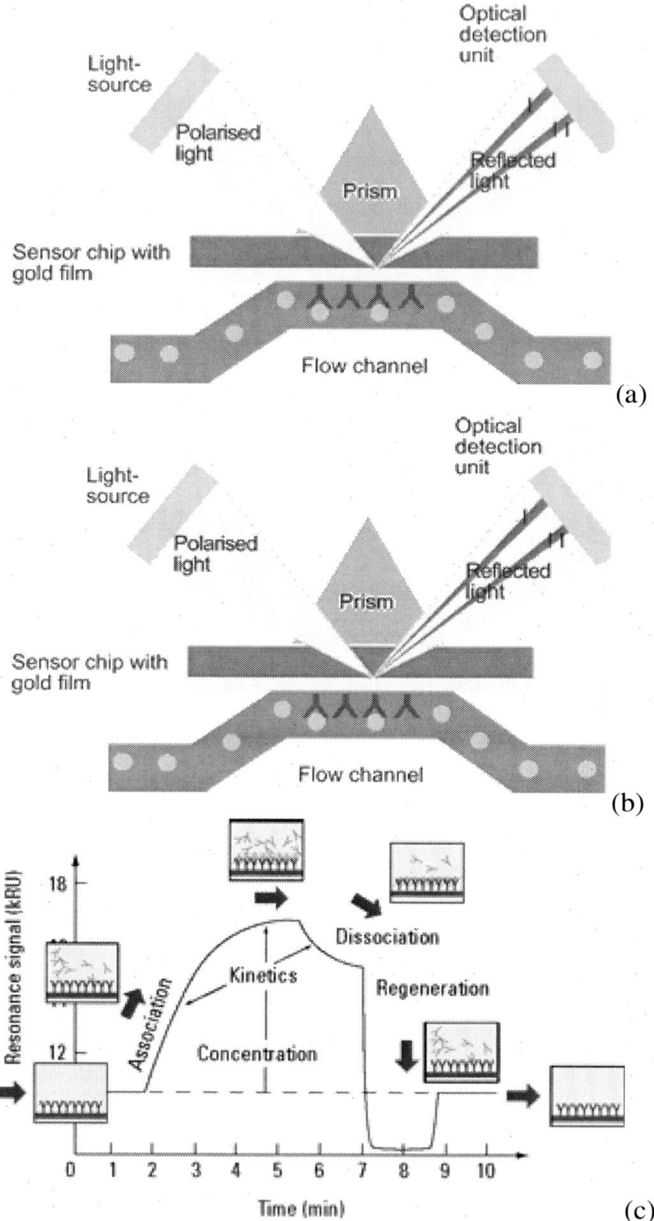

Figure 4a, b, c. Step by step working of SPR (see text for description). (Source & Citation: Markey, F. (1999). *BIA journal 1*, 14-17).

1.2.1. Binding Response Curves

The increase in mass associated with a binding event causes a proportional increase in refractive index which is observed as a change in response (Figure 5). This response is defined in Response Units (RU) where 1 RU is equivalent to 1μ Refractive index unit, which represents approximately 1pg of protein/mm^2. During an injection, the injected analyte (A_0) is transported to the interaction surface by convection and diffusion where its concentration (A) may change due to complex formation. This change in concentration is minimized by ensuring a high mass transport coefficient (km) hence the use of nanoliter (nL) scale flow cells. Immobilizing a low concentration of ligand is also extremely effective in ensuring that mass transport of analyte does not become limiting. The response increase during injection of analyte (A) is due to the formation of analyte-ligand complexes (AB) at the surface. The decrease in response after analyte injection is due to the dissociation of the complexes. A binding interaction model may be fitted to this data allowing the kinetic and/or affinity constants to be determined. If particularly fast interactions are recorded, a model that compensates for mass transport limitation (i.e. two compartment model) may be fitted. Bulk index effects and non-specific binding are removed by real-time reference curve subtraction [8,9].

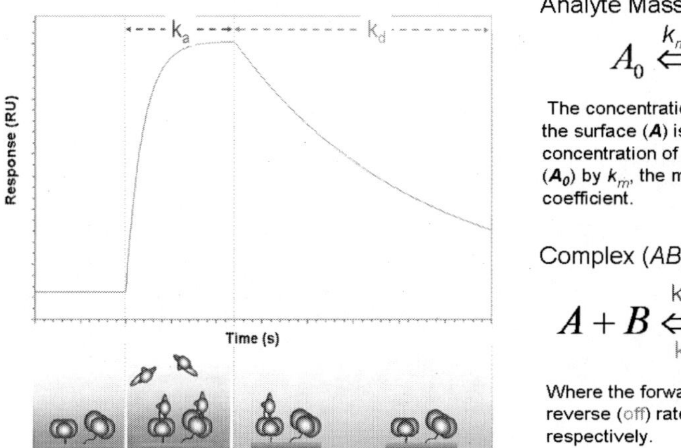

Figure 5. Interpretation of Binding Response Curves.
(http://www.sensiqtech.com/uploads/file/support/spr/Overview_of_SPR.pdf).

Realization

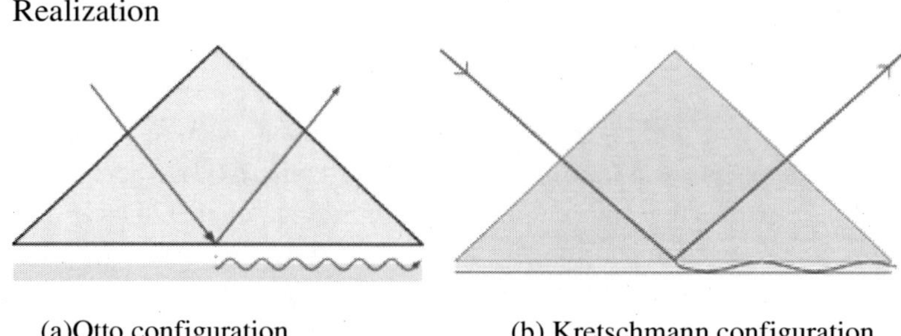

(a)Otto configuration (b) Kretschmann configuration

Figure 6. Otto configuration; (b) Kretschmann configuration. (Source & Citation: Zeng, S., Yu, X., et al. (2012). *Sensors and Actuators B: Chemical* 176: 1128. doi:10.1016/j.snb. 2012.09.073).

Chapter 2

PRINCIPLES AND MECHANISM BEHIND SPR

2.1. PHYSICAL BASIS OF SPR

Surface plasmons, also known as surface plasmon polaritons, are surface electromagnetic waves that propagate in a direction parallel to the metal/dielectric (or metal/vacuum) interface. These are charge-density oscillations localized at the interface between a metallic film (active layer) and dielectric surface. Since the wave is on the boundary of the metal and the external medium (air or water for example), these oscillations are very sensitive to any change of this boundary, such as the adsorption of molecules to the metal surface.

To describe the existence and properties of surface plasmons, one can choose from various models (quantum theory, Drude model, etc.). The simplest way to approach the problem is to treat each material as a homogeneous continuum, described by a frequency-dependent relative permittivity between the external medium and the surface. This quantity, hereafter referred to as the materials' "dielectric constant," is complex-permittivity. In order for the terms which describe the electronic surface plasmons to exist, the real part of the dielectric constant of the metal must be negative and its magnitude must be greater than that of the dielectric. This condition is met in the IR-visible wavelength region for air/metal and water/metal interfaces (where the real dielectric constant of a metal is negative and that of air or water is positive) [8,9].

2.2. LOCALIZED SURFACE PLASMON RESONANCE (LSPR)

SPR in nanometer-sized structures is called *Localized surface plasmon resonance*. LSPRs are collective electron charge oscillations in metallic nano-particles that are excited by light. They exhibit enhanced near-field amplitude at the resonance wavelength. This field is highly localized at the nanoparticle and decays rapidly away from the nano-particle/dieletric interface into the dielectric background, though far-field scattering by the particle is also enhanced by the resonance. Light intensity enhancement is a very important aspect of LSPRs and localization means the LSPR has very high spatial resolution (sub wavelength), limited only by the size of nano-particles. Because of the enhanced field amplitude, effects that depend on the amplitude such as magneto-optical effect are also enhanced by LSPRs [10,11].

In order to excite surface plasmons in a resonant manner, one can use an electron or light beam (visible and infrared are typical). The incoming beam has to match its impulse to that of the plasmon. In case of p-polarized light (polarization occurs parallel to the plane of incidence), this is possible by passing the light through a block of glass to increase the wavenumber (and the impulse), and achieve the resonance at a given wavelength and angle. S-polarized light (polarization occurs perpendicular to the plane of incidence) cannot excite electronic surface plasmons. Electronic and magnetic surface plasmons obey the following dispersion relation:

$$K(\omega) = \frac{\omega}{c}\sqrt{\frac{\varepsilon_1\varepsilon_2\mu_1\mu_2}{\varepsilon_1\mu_1 + \varepsilon_2\mu_2}}$$

Where, ϵ is the dielectric constant, and μ is the magnetic permeability of the material (1: the glass block, 2: the metal film).

There are various methods that use different coupling devices between the incident light and surface Plasmon. When using light to excite SP waves, there are two configurations which are well known. In the Otto setup, the light illuminates the wall of a glass block, typically a prism, and is totally internally reflected. A thin metal film (for example gold) is positioned close enough to the prism wall so that an evanescent wave can interact with the plasma waves on the surface and hence excite the plasmons.

The mostly widely used method is the Kretschmann geometry where the metallic film is coated directly on the prism or on a plate held on a prism with an appropriate index matching liquid. In the Kretschmann configuration, the metal film is evaporated onto the glass block. The light again illuminates the glass block, and an evanescent wave penetrates through the metal film. The plasmons are excited at the outer side of the film. This configuration is used in most practical applications (Figure 6).

2.3. SPR EMISSION

When the surface plasmon wave interacts with a local particle or irregularity, such as a rough surface, part of the energy can be re-emitted as light. This emitted light can be detected behind the metal film from various directions.

So far, SPR has been exploited only in the visible and the near-infrared region due to the wide availability of optical materials and efficient light sources and detectors. Nevertheless, most of the optical absorption spectral lines associated with the vibrational frequencies of gas molecule takes place in the mid infrared region of the optical spectrum. These absorption characteristics allow designing selective gas sensors using thermoelectric, pyroelectric or photoaccoustic detectors in the mid infrared range [12,13,14].

Chapter 3

INSTRUMENTS BASED ON SPR PHENOMENA

3.1. HISTORICAL OVERVIEW

The development of [10,11] first biosensor based on surface plasmon resonance (henceforth referred to as SPR), has triggered the use of this technique for studying the binding interactions. With the popularization of the technique, several SPR-based systems [15,16,17] came into existence but the one provided by BIAcore is by far the most widely used one [10,11]. BIAcore produced by BIAcore AB has developed a range of instruments, i.e. BIAcore X, 2000 and 3000 series with veritable options that include choice with automation, flow cell number, volume, time resolution, refractive index range, sample recovery etc. Presently over 2000 publications had reported results obtained using the BIAcore. The data could have skewed towards higher side, had it not been to the cost and the pitfalls associated with obtaining accurate quantitative data [17,18,19,20]. The latter has discouraged many investigators and led to the perception that the technique might be flawed, which is not correct because the pitfalls are common to many binding techniques and, once understood, they are easily avoided [16,20,21]. Furthermore, BIAcore and many other SPR based instruments are sensitive enough to detect and analyse weak macromolecular interactions, allowing measurements that are not possible using any other technique [16,22]. This Chapter intends to throw a light on the usage of SPR, with an emphasis on avoiding pitfalls. Tedious literature on the routine operation and maintenance of the SPR instruments are avoided as this is comprehensively described in the instrument manual.

3.2. Principles and Mechanism behind Working of SPR Based Instruments

The physical principles underlying the working of SPR are complicated, though an adequate working knowledge of the technique does not entail a detailed theoretical understanding of the technique. SPR-based instruments use an optical method to measure the refractive index close to (within ~300 nm) a sensor surface. This surface forms the floor of a small flow cell, usually 20 to 60 nL in volume, through which a running buffer passes under continuous flow at the rate of 1 to 100 $\mu Lmin^{-1}$. To detect an interaction one molecule which is called the ligand is immobilised onto the sensor surface and subsequently its binding partner which is also called the analyte is injected in aqueous solution (in sample buffer) through the flow cell, also under continuous flow. As the analyte binds to the ligand the accumulation of protein on the sensor surface results in an increase in the refractive index. This change in the refractive index is quantitated in real time, and the results are plotted as Response / Resonance units (RUs) versus time. The resulting plot is called a Sensorgram. Importantly, a 'Background Response' will also be generated if running and sample buffers differ in the refractive indices. Therefore, this background response needs to be subtracted from the sensorgram and that will give the actual binding response. To get the background response, analyte is injected through a control or reference flow cell, which has no ligand or an irrelevant ligand immobilized to the sensor surface [2].

One RU represents the binding of approximately 1 picogram (pg) protein/mm^2. Usually greater than 50 pg/mm^2 of analyte binding is required for meaningful interpretation. It is practically very difficult to immobilise a sufficiently high density of ligand onto a surface to achieve this level of analyte binding. To overcome this, usually a sensor surface with a 100 to 200 nm thick carboxymethylated dextran matrix is attached. Much higher levels of ligand immobilisation are possible by effectively adding a third dimension to the surface. Having very high levels of ligand however, too has some drawbacks. First limitation is that the mass-transport (analyte delivered to the surface) becomes the rate-limiting step as with such a high ligand density the rate at which the surface binds the analyte may exceed the rate at which the analyte can be delivered to the surface. In this situation, the measured association rate constant (k_{on}) will be slower than the true k_{on}. Second problem is related to the Rebinding i.e. after dissociation of the analyte, it can rebind to the unoccupied ligand before diffusing out of the matrix and being washed from the flow cell. As a result, the measured dissociation rate constant (apparent k_{off}) will be slower than the true k_{off}. Dextran matrix may compound these kinetic artefacts (mass transport limitations and re-binding) and can affect all surface-binding techniques [2].

Chapter 4

APPLICATIONS

In addition to the obvious use of SPR to quantitate the binding interactions, this section also outlines some applications for which it is probably not the technique of choice. Technical improvisations however have the potential of extending the range of applications in future for which the SPR is useful.

4.1. OPTICAL SENSOR BASED ON SPR OPERATING IN THE MID-INFRARED RANGE

The experimental set up is based on Kretschmann geometry with Ti/Au layers deposited on a CaF_2 prism where light excitation is provided by a Quantum cascade laser (QCL) source. Evidence of SPR is presented and the sensing capability demonstrated by using CO_2 and N_2 mixtures as test samples. Due to absorption of CO_2 at this wavelength, it is shown that the sensitivity of this configuration is five times higher than a similar SPR sensor operating in the visible range of the spectrum [24]

Surface plasmons have been used to enhance the surface sensitivity of several spectroscopic measurements including fluorescence, Raman scattering, and second harmonic generation. However, in its simplest form, SPR reflectivity measurements can be used to detect molecular adsorption, such as polymers, DNA or proteins, etc. Technically, it is common, that the angle of the reflection minimum (absorption maximum) is measured. This angle changes in the order of 0.1° during thin (about nm thickness) film adsorption. In other cases the changes in the absorption wavelength is followed [15]. The mechanism of detection is based on the fact that, adsorbing molecules cause changes in the local index of refraction, changing the resonance conditions of the surface plasmon waves (Figure 7a & b).

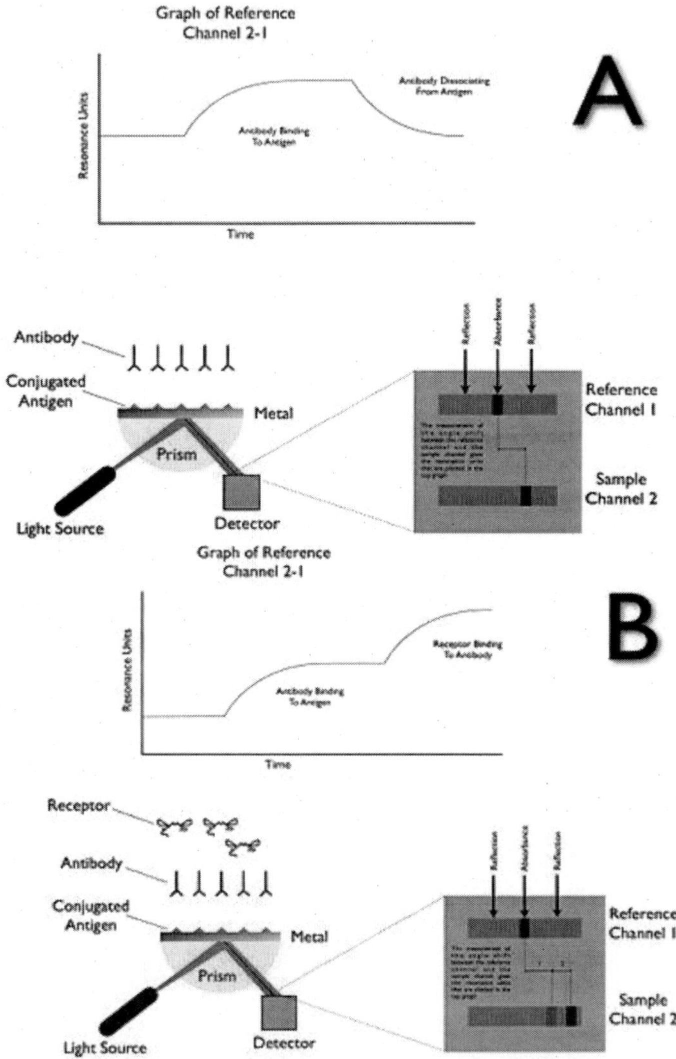

Figure 7a&b. Diagram summarizes the data displayed by a Surface Plasmon Resonance (SPR) machine using a capturing molecule followed by a ligand. (Source & Citation: Sabban, S (2011). PhD thesis, The University of Sheffield.

If the surface is patterned with different biopolymers, using adequate optics and imaging sensors (i.e. a camera), the technique can be extended to surface plasmon resonance imaging (SPRI). This method provides a high contrast of the images based on the adsorbed amount of molecules, somewhat similar to Brewster angle microscopy (latter is most commonly used together with a Langmuir-Blodgett trough).

For nano-particles, localized surface plasmon oscillations can give rise to the intense colors of suspensions or sols containing the nano-particles. Nano-particles or nano-wires of noble metals exhibit strong absorption bands in the ultraviolet-visible light regime that are not present in the bulk metal. This extraordinary absorption increase has been exploited to increase light absorption in photovoltaic cells by depositing metal nano-particles on the cell surface [16]. The energy (color) of this absorption differs when the light is polarized along or perpendicular to the nano-wire [17]. Shifts in this resonance due to changes in the local index of refraction upon adsorption to the nano-particles can also be used to detect biopolymers such as DNA or proteins. Related complementary techniques include plasmon waveguide resonance, QCM, extraordinary optical transmission, and dual polarization interferometry (Figure 7a & b).

4.2. ADVANTAGES OF SPR

Evaluation of Macromolecules

SPR is particularly suited to evaluate the binding of recombinant proteins to natural ligands and mAbs. Setting up an assay for any particular protein is very fast, and the data provided are highly informative. This is particularly important as most laboratories studying biological problems at the molecular or cellular level work with recombinant proteins. To confirm, whether recombinant protein has the same structure as its native counterpart, can be detected by the fact that with the possible exception of enzymes, protein binds its natural ligands. Such interactions involve multiple residues, which are far apart in the primary amino acid sequence, hence require a correctly folded protein. In the absence of natural ligands monoclonal antibodies (mAbs) that are known to bind to the native protein are an excellent means of assessing the structural homology and integrity of the recombinant protein [23].

4.3. Binding kinetics of a Model Antibody-antigen System, HSA Binding to Anti-HSA IgG. Anti-HSA Is Biotinylated with Approximately 6 Biotin Groups and Captured on a Planar Neutr-Avidin Sensor Slide

The SR7000DC monitors this antibody-antigen interaction in real-time with simultaneous monitoring of sample and reference channels. The experimental conditions for this assay are summarized in table 1.

Table 1. The experimental conditions for this assay are summarized below

Ligand	**Analyte**	**Analyte Concentrations**	**Ligand Association Time**	**Dissociation Time**	**Regeneration Solution**
Anti-HSA 1	HSA	80, 40, 20, 10, 5, 2.5, 1.25 nM	3 min	5 min	Cocktail

Source & Citation: Anderson, K., Hamalainen, M. & Malmqvist, M. (1999). *Anal Chem 71*, 2475-2481.

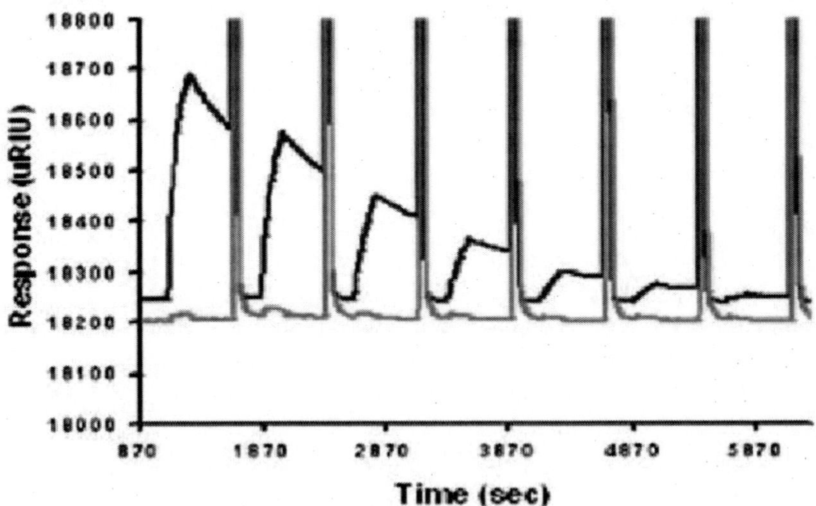

Figure 8a. SPR sensorgram showing injections of HSA at concentrations of 80, 40, 20, 10, 5, 2.5, and 1.25 nM. (Source & Citation: Anderson, K., Hamalainen, M. & Malmqvist, M. (1999). *Anal Chem 71*, 2475-2481).

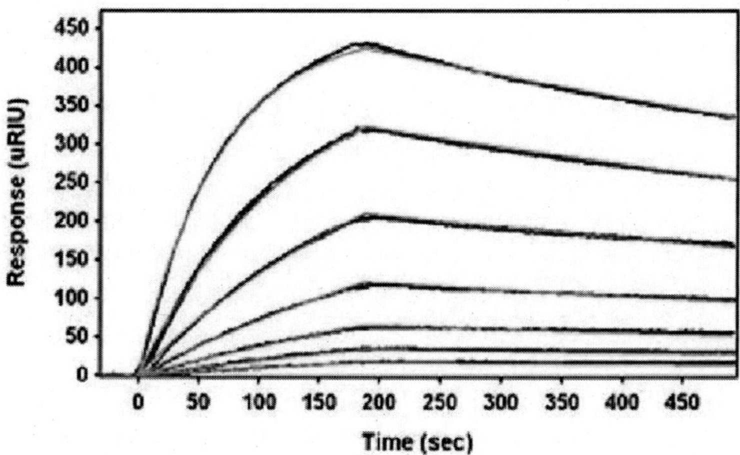

Figure 8b. Normalized response versus time plots of HSA binding to anti-HSA fit to a simple bimolecular model in Scrubber (Source & Citation: Anderson, K., Hamalainen, M. & Malmqvist, M. (1999). *Anal Chem 71*, 2475-2481).

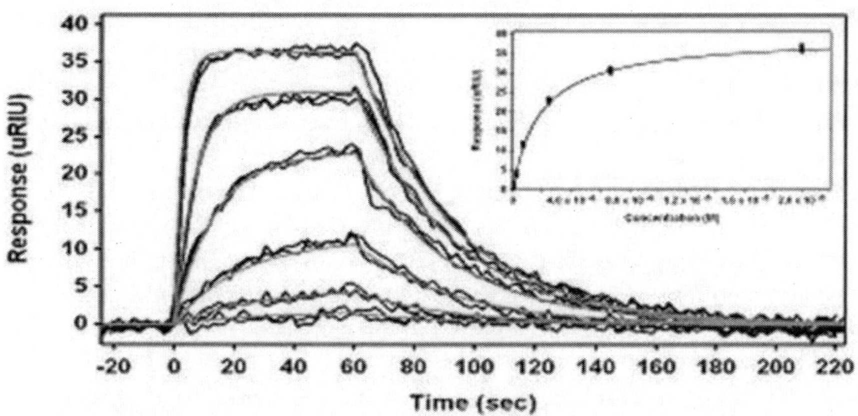

Figure 9. Normalized response versus time plots of 4-CBS binding to CAII fit (red lines) to a simple bimolecular model in Scrubber. The inset is the corresponding Langmuir binding isotherm (Source & Citation : Papalia, G.A., Leavitt, S., et al. (2006). *Anal Biochem 359*, 94-105).

Figure 8a shows raw data from a cycle of HSA injections across the two channels; one being the sample side (with captured anti-HSA) and the other serving as a reference (surface without anti-HSA). The results show that the binding is highly specific with very minimal nonspecific binding on the reference channel.

Figure 8b shows the normalized HSA binding curves along with the fit (red lines) to a simple bimolecular model using Scrubber (Biologic Software). Each concentration is injected at least twice to verify reproducibility. The association rate constant (k_a) is found to be 1.66e5 M^{-1s-1} and the dissociation rate constant (k_d) is determined to be 7.70$e^{-4\ s-1}$. Thus, the equilibrium dissociation constant (K_D) (i.e., the ratio of kd to ka) is 4.65 nM for this interaction [25].

4.4. SPR Binding Experiment between CAII and an Inhibitor, 4-Crboxybenzenesulfonamide (4-CBS); A Small Molecule with a Molecular Weight of 201 Da

The kinetic data is fit to a simple bimolecular model using Scrubber (Biologic Software) (red lines) and the equilibrium data in the inset of Figure 9 is fit to a Langmuir binding isotherm model (solid line). This small molecule binding experiment was carried out on four separate occasions and Table 2 summarizes assay condition while, Table 3 presents the association and dissociation rate constants determined from fits to a simple bimolecular model in Scrubber along with the equilibrium dissociation constants (K_D) calculated from the kinetic and equilibrium data, respectively. The results show that the system is highly reproducible and the K_D values determined through the kinetic and equilibrium analysis are in very good agreement with each other and correlate very well with that reported in the literature [26, 27].

Table 2. The experimental conditions for this assay are summarized below

Ligand	Analyte	Analyte Concentrations	Ligand Association Time	Dissociation Time	Regeneration Solution
CAII 1	4-CBS	20, 6.7, 2.2, 0.74, 0.25, 0.082 μM	1 min	3 min	None

Source & Citation: Paplia, G.A., Leavitt, S., et al., (2006), Anal Biol.

Table 3. Summary of results from four separate experiments

	k_a (M-1s-1)	k_d (s-1)	K_D (μM) (Kinetic)	K_D (μM) (Equilibrium)
Run 1	2.9e4	2.9e-2	0.96	1.2
Run 2	2.8e4	2.9e-2	1.03	1.2
Run 3	3.2e4	3.2e-2	1.01	1.1
Run 4	3.6e4	4.2e-2	1.19	1.2

Source & Citation: Paplia, G.A., Leavitt, S., et al., (2006), Anal Biol.

4.5. Thermodynamic Investigation of an Enzyme-Inhibitor Pair

Understanding the thermodynamics of a system gives valuable insight into the mechanism and forces leading to an interaction. Reichert's SR7000DC system provides the opportunity to acquire thermodynamic data for molecular interactions of interest owing to its precise temperature control from 10°C below ambient to 90°C and was utilized to characterize the thermodynamics of a small molecule inhibitor-enzyme pair, specifically carbonic anhydrase II (CAII) and 4-carboxybenzenesulfonamide (4-CBS in the concentrations; 20, 6.7, 2.2, 0.74, 0.25, 0.082 μM). The binding constants of this interaction are measured with respect to temperature ranges (15, 25, 30, 35°C) to thermodynamically profile this interaction. As shown in figures 10 and 11, temperature has a drastic effect on the profile of the response curves. Specifically, the association and dissociation rate constants increase with temperature. To quantify the change in rates, the data at each temperature was fit to a simple bimolecular model using Scrubber (Biologic Software) to determine the rate constants and the equilibrium dissociation rate constants.

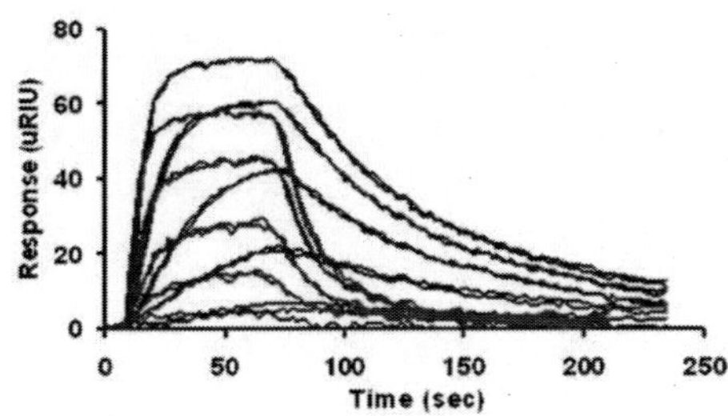

Figure 10. Temperature dependent response curves of 4-CBS binding to CAII at 20°C (blue lines) and 35°C (red lines).

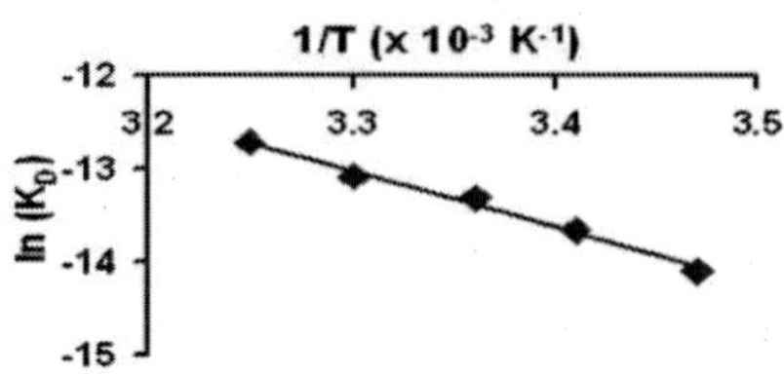

Figure 11. van't Hoff analysis of the data showing binding of 4-CBS with CAII by plotting ln *KD* versus *1/T*. The data fits fairly well to a linear regression model ($r^2 = 0.989$), thus the thermodynamic parameter ΔH can be determined directly from the non-integrated form of the van't Hoff equation. In this case, the slope is ΔH/R. Thus, ΔH is determined to be -6.0 kcal/mol for this inhibitor-enzyme pair.

4.6. Small Volume Injections with SR7500 Syringe Pump

The Reichert SR7500 syringe pump includes a feature for small volume injections referred to as recycle mode. The recycle mode is ideal when sample volume is limited. This unique feature is typically used either for immobilization of ligand or to obtain equilibrium or kinetic data from analyte injections when only minimal sample volume is available for injection (1 - 50 µL).

A small volume of sample can be aspirated by the pump and then recycled to conserve on sample. To accomplish this, a short piece of tubing is connected to the flow cell outlet (4" long x 1/16" OD x 0.005" ID, volume = 1.3 µL). This tubing is submerged directly into the sample container (e.g., a micro centrifuge tube) and the sample is aspirated into the flow cell at a particular rate. The sample is then pumped back and forth in the flow cell at a defined recycle rate (Figure 12a, b). The range of available volumes and rates are summarized in table 4.

Table 4. The range of available volumes and rates

Aspirate Volumes	Aspirate Rates	Recycle Volumes	Recycle Rates
5, 10, 15, 20, 30, 40, 50 µl	1, 2, 5, 10 µl	1-10 µl	25, 50, 100, 200, 500 to 6000 in 500 increments (µl)

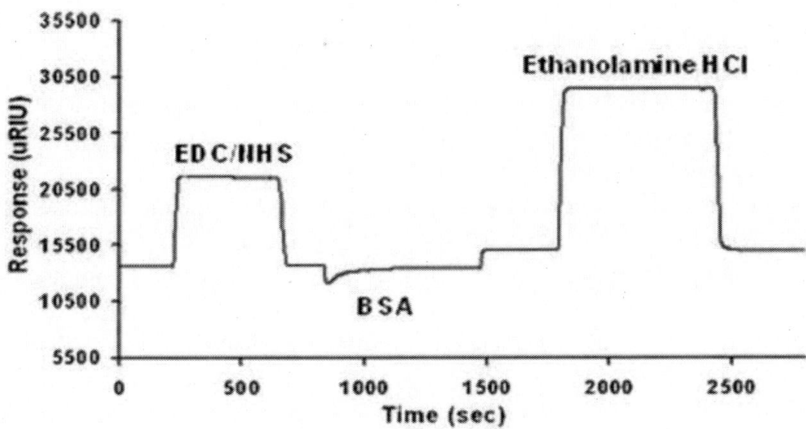

Figure 12a. Immobilization of Bovine serum albumin (BSA). Specifically, BSA is amine coupled to a planar mixed alkanethiol surface. A sample volume of 10 μL of BSA is aspirated at a rate of 10 μL/min into the flow cell. The BSA sample (recycle volume of 1 μl) is then pumped forward and reverse at a recycle rate of 500 μl/min. The immobilization of BSA was carried out for 630 sec.

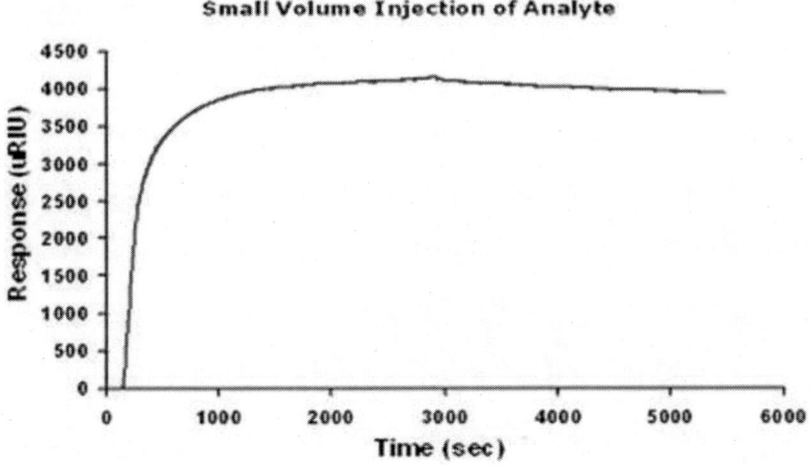

Figure 12b. Binding of anti-Bovine serum albumin (Anti-BSA) to BSA despite an extremely low sample volume. Specifically, a sample volume of 15 μL of anti-BSA is aspirated at a rate of 10 μL/min into the flow cell, and then a recycle volume of 1 μL is used at a recycle rate of 500 μL/min for the association binding curve. The association time is 2800 sec and the response approaches equilibrium.

4.7. Using Combined Electrochemistry and SPR to Monitor the Electro-polymerization of Aniline

Conducting polymer thin films have been the focus of increased investigation because of their importance to a variety of applications including displays, batteries and sensors. Polyaniline (PAn) is a conducting polymer that has been widely studied because of its unique properties. These features include its relative ease of synthesis, remarkable environmental and chemical stability, and the drastic change in its electrical conductivity upon simple oxidation and reduction. Owing to the SR7000DC's open architecture, Reichert offers specialized flow cells/wells for performing other measurements in combination with surface plasmon resonance (SPR). Reichert SR7000DC system (Figures 13, 14) presents the combination of electrochemistry and SPR to monitor the growth of a PAn film on the surface of bare gold. Specifically, cyclic voltammetry (CV) is used as the deposition method while SPR simultaneously monitors the growth of the PAn film in real time [28, 29]. The experimental conditions for this assay are summarized in Table 5.

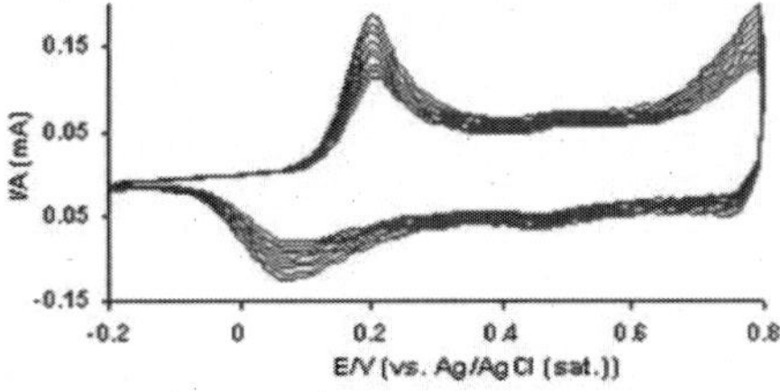

Figure 13. Cyclic voltammograms of the electropolymerization of aniline in 0.5 M H_2SO_4 on the gold electrode (Source & Citation: Baba, A., Tian, S. et al. (2004). *J Electroanal Chem 562*, 95-103).

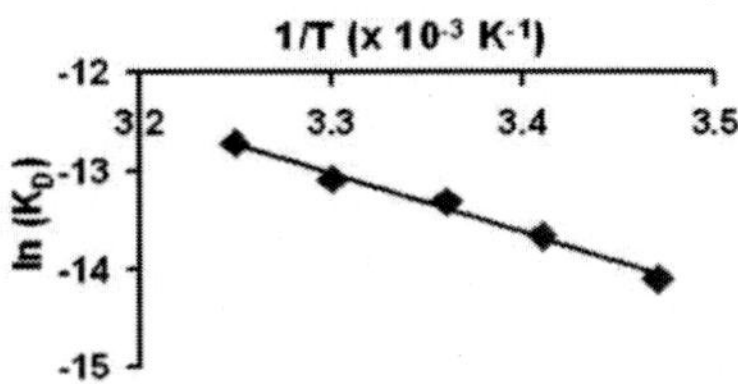

Figure 14. SPR response during the electropolymerization of aniline. The figure shows the SPR reflectivity data taken from each set of 10 potential cycles (Source & Citation: Baba, A., Tian, S. et al. (2004). *J Electroanal Chem 562*, 95-103).

Table 5. The experimental conditions for this assay are summarized below

Aniline Concentration	Running Solution CV	Scan Rate CV Scan	Potential Range
0.1 M	0.5 M H2SO4	100 mV/s	-0.2 to 0.8 V

Source & Citation : Baba, A. , Tian, S. et al. (2004). *J Electroanal Chem 562*, 95-103.

4.8. Anisotropic Surface Plasmon Resonance Imaging Biosensor

SPR is a physical phenomenon that occurs when light is coupled to a thin layer of metal usually silver or gold under certain conditions (specific angle of incidence, wavelength, polarization, and metal thickness). An evanescent plasmon wave propagates along the metal-dielectric interface (see Figure 15a). Changes in either the refractive index or biomolecular layer thickness in the vicinity of the metallic interface can be monitored as a shift in the resonance curve. SPRI allows direct measurement of the optical constants of the dielectric layer above the metal. This gives an indirect measure of the increase in mass resulting from adsorption of biomolecular targets.

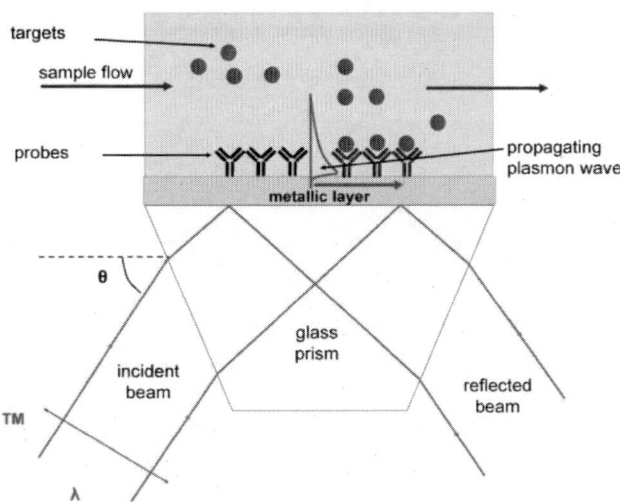

Figure 15a. Diagram of a typical surface plasmon resonance imaging (SPRI) system. The transverse magnetic (TM)–polarized incident beam of wavelength λ is refracted through a high-index glass prism of angle θ (Source & Citation: Duval, A., Aide, A., et al., (2007). *SPIE* 2-3, doi: 10.1117/2.1200710.0882).

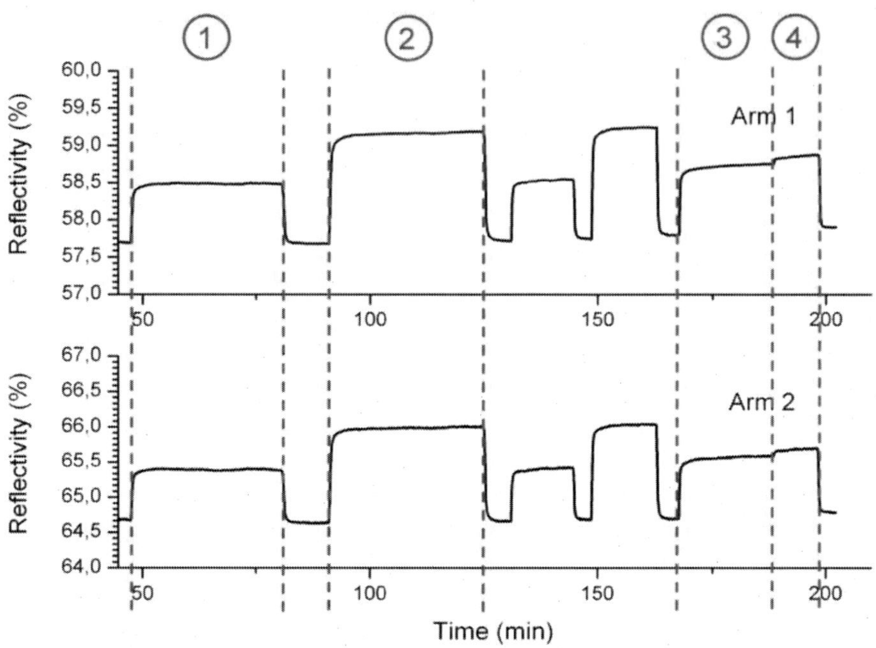

Figure 15b. Evolution of the average reflectivity for each arm when sequentially injecting (1) 1% glycerol in water, (2) 2% glycerol in water (repeated two times), and finally (3) 1× PBS (phosphate-buffered saline), then (4) 1.25× PBS (Source & Citation: Duval, A., Aide, A., et al., (2007). *SPIE* 2-3, doi: 10.1117/2.1200710.0882).

SPRI enables us to evaluate precisely, in real time, the adsorption and desorption kinetics of each spot throughout the chip's surface (2D array). Moreover, additional dimensions can be exploited, such as angle of incidence, wavelength, and polarization. This dynamic information provides a multidimensional `image' of the interactions [30].

4.8.1. Anisotropic SPRI: Extracting Conformational Parameters

Changes in the shape of biomolecules are often important in analyzing biological processes. For example, different conformations of a protein could result in different functions, or in disease. The system is based on two orthogonal SPRI-sensing arms. Each arm is able to retrieve real-time reflectivity data about every spot in the 2D array. We can then perform a differential measurement and extract the average anisotropy—i.e., how physical properties vary with direction—at each point of the array. With this information, we can recover shape and directional features for the biomolecular layer (with known optical constants), such as order parameters or the average orientation of the biomolecules tethered to the chip [30].

Determining anisotropy from two sensing arms implies that both give similar results accurately and consistently for isotropic samples. To verify the stability of the setup, an experiment was conducted using a bare gold chip covered with isotropic mixtures of different refractive indexes laid down one after the other. Two cameras recorded the reflectivity information at each location on the surface during the course of the experiment (see Figure 15b). Results were found to be identical within a 3.10−2% differential noise limit [15].

4.8.2. Future Perspectives

Over the short term, focus will be on validating the SPRI system for anisotropic samples. For instance, binding magnetic beads to the biomolecules on the chip's gold surface and to record changes in their orientation induced by a magnetic field. This type of system will be useful for detecting local flow differences within the same sample, enabling more accurate calibration of the results. In the long term, the conformational data will aid in discriminating variants of proteins. This is important, for example, in prion-type diseases, which are caused by a mutated protein in which only a handful of atomic bonds have changed orientation.

4.9. Electrostatic / Electrochemical SPR

In situ optical surface plasmon resonance spectroscopy can be used to monitor hybridization kinetics for unlabeled DNA in tethered monolayer nucleic acid films on gold in the presence of an applied electrostatic field. This can investigate the effect of simple electrostatic charging on the interactions of surface-bound monolayer nucleic acid films with unlabeled DNA target oligonucleotides by using a combination of electrochemical control and *in situ* real-time surface plasmon resonance (SPR) spectroscopy detection. The monolayer DNA thiol films used in this work are tethered directly to the SPR metal sensor surface through a gold–thiol covalent attachment. In this, a repulsive potential preferentially denatures mismatched DNA hybrids within a few minutes, while leaving the fully complementary hybrids largely intact. This sequence selectivity in hybrid denaturation imparts an extremely high stringency for assaying DNA interactions and represents an extremely simple method for mutation detection based purely on electrostatic charging effects. This also has the advantage of using label-free SPR detection of hybridization and denaturation by monitoring gain or loss of DNA at the interface in the presence of applied electrochemical field.

Increasing research efforts are directed toward the detection of nucleic acid interactions with immobilized oligonucleotide probes for DNA microarray applications. Limitations of most current technologies include complex DNA immobilization procedures, the need for fluorescence or other labeling, and slow hybridization kinetics that require long incubation times. In addition, those experimental conditions that optimize DNA duplex formation often also reduce the stringency of hybridization. In light of these, ESPR offers and opens new avenues for nucleic acid hybridization chemistry for DNA microarray applications [31].

4.10. SPR for Detecting Single Nucleotide Polymorphism (SNP)

Miura et al. (2008) reported a SPR sensor carrying an organic ligand for the detection of single nucleotide polymorphism (SNP). It is based on the following principle; selected aromatic ligands can bind to a nucleobase opposite to an abasic site (AP-site) in DNA duplexes. An AP site containing probe DNA is hybridized to a target DNA. Fluorescence signaling is obtained by direct interaction between ligands and target nucleobase and SNP genotype of samples can be distinguished by the combination of ligands with selectivity for target nucleobases. This system was further developed for SPR assay by immobilizing an AP-site binding ligand (3,5 diaminopyrazine) on the surface of the sensor chip with a high selectivity and sensitivity for thymine in AP site containing DNA duplexes. Such system was presumed to be developed for gene detection chemistry based on DNA binding small molecules [32].

4.11. LOCALISED SURFACE PLASMON RESONANCE (LSPR)

Localised surface plasmon is charge density oscillations confined to metallic nano-particles and metallic nano structures. LSPR can be utilized as nano-scale optical devices and sensors. Nano-scale optical and photonic devices based on LSPR are high quality parabolic mirrors, nano scale optical coupling, Plasmon waveguides, optical transmission through sub-wavelength holes, optical data storage, optical information transfer below the diffraction limit.

Assays based on LSPR are, ligand recognition by mAb, Lectin, Human Serum Albumin, Streptavidin. Additionally, there can be assays based on aggregation of metallic nano-particles. Labelling can also be done by metallic nano-particles using LSPR. Signal amplification can also be done by LSPR. Exploitation of LSPR as nano-scale optical devices and sensors has seen an unprecedented growth. Further understanding and greater cooperativity between physicist, chemist and material scientist can take its utility to a new level [33].

4.12. MAGNETIC PLASMON RESONANCE

Recently, there has been an interest in magnetic surface plasmons. These require materials with large negative magnetic permeability, a property that has only recently been made available with the construction of meta-materials.

4.13. EQUILIBRIUM MEASUREMENTS (AFFINITY AND ENTHALPY)

Equilibrium analysis is very time consuming and requires multiple sequential injections of analyte at different concentrations (and at different temperatures). Therefore, it is only practical to perform equilibrium analysis on interactions that attain equilibrium within about 30 min. The time to reach equilibrium is determined by the dissociation rate constant or k_{off}; a useful rule of thumb is that an interaction should reach 99% of the equilibrium level within $4.6/k_{off}$ seconds. Usually high affinity interactions ($K_D < 10$ nM) have very slow k_{off} values and are therefore unsuitable for equilibrium analysis.

Conversely, very weak interactions (K_D >100 μM) are easily studied. The way out is to use a concentrated sample in a small volume. The small sample volumes required for SPR injections (<20 μL) make it feasible to inject the very high concentrations (>500 μM) of protein required to saturate low affinity interactions [34].

Equilibrium affinity measurements via SPR are highly reproducible. This feature coupled with precise temperature control makes it possible to estimate binding enthalpy by van't Hoff analysis [35]. This involves measuring the (often small) change in affinity with temperature. Although, this is not as rigorous as calorimetry, the protein required is much less [23].

4.14. Kinetic Measurements

SPR based instruments are user-friendly and it is easy to generate kinetic data using it. Obtaining accurate kinetic data requires a thorough understanding of binding kinetics and source of artefacts (source of artefacts). These instruments generate real-time binding data, which makes it well suited to the analysis of binding kinetics. However, there are important limitations to the kinetic analysis. Largely because of mass transport limitations due to which it is difficult to measure accurately k_{on} values faster than about 10^6 $M^{-1}s^{-1}$. The upper limit is dependent on the size of the analyte. Analytes with a greater molecular mass facilitates faster k_{on} values because the larger signal produced by a large analyte allows the experiment to be performed at lower ligand densities, and lower ligand densities require lower rates of mass transport. For different reasons measuring k_{off} values slower than 10^{-5} s^{-1} or faster than ~1 s^{-1} is difficult [23].

4.15. Analysis of Mutant Proteins

SPR is very convenient for analysing mutants generated by site-directed mutagenesis [36, 37]. Mutants can be expressed as tagged proteins by transient transfection and then captured from crude tissue culture supernatant using an antibody to the tag, thus effectively purifying the mutant protein on the sensor surface. SPR helps to visualise the capture of proteins from crude mixtures onto the sensor surface. This elucidates the effect of the mutation on the binding properties (affinity, kinetics, and even thermodynamics) of the immobilised protein. This is the only practical way of quantifying the effect of mutations on the thermodynamics and kinetics of weak protein/ligand interactions [38].

4.16. Limitations of SPR

4.16.1. High Throughput Assays

SPR is unsuitable for high throughput assays as only one sample can be analysed at a time, with each analysis taking 5-15 min. Automation is of no use as well because the sensor surface deteriorates over time and with re-use. Microfluidic system getting blockages or air bubbles are common in long experiments, particularly when many samples are injected.

4.16.2. Concentration Assays

Concentration measurements cannot be done on SPR because these require the analysis of many samples in parallel, including the standard curve. Also, for optimal sensitivity, concentration assays require long equilibration periods.

4.16.3. Studying Small Analytes

SPR measures the mass of material binding to the sensor surface, so, very small analytes (M_r <1000) give very small responses. However, recent improvements in signal to noise ratio have made it possible to measure binding of such small analytes. A very high surface concentration of active immobilised ligand (~1 mM) is needed, which is difficult to achieve. Also, at such high ligand densities accurate kinetic analysis is not possible because of the dual limitations of mass-transport and re-binding. Therefore, only equilibrium analysis is possible with very small analytes and that too under optimal conditions. This assessment needs relook as and when future improvements are made in the signal to noise ratio.

Chapter 5

DATA INTERPRETATION

5.1. FRESNEL FORMULA

The most common data interpretation is based on the Fresnel formulas, which treat the formed thin films as infinite, continuous dielectric layers. This interpretation may result multiple possible refractive index and thickness values. However, usually only one solution is within the reasonable data range [39, 40].

Metal particle plasmons are usually modeled using the Mie scattering theory [41, 42].

In many cases no detailed models are applied, but the sensors are calibrated for the specific application, and used with interpolation within the calibration curve.

One of the first common applications of surface plasmon resonance spectroscopy was the measurement of the thickness (and refractive index) of adsorbed self-assembled nano-films on gold substrates. The resonance curves shift to higher angles as the thickness of the adsorbed film increases. This is a 'static SPR' measurement.

When higher speed observation is desired, one can select an angle right below the resonance point (the angle of minimum reflectance), and measure the reflectivity changes at that point. This is the so called 'dynamic SPR' measurement. The interpretation of the data assumes that the structure of the film does not change significantly during the measurement (Figure 16).

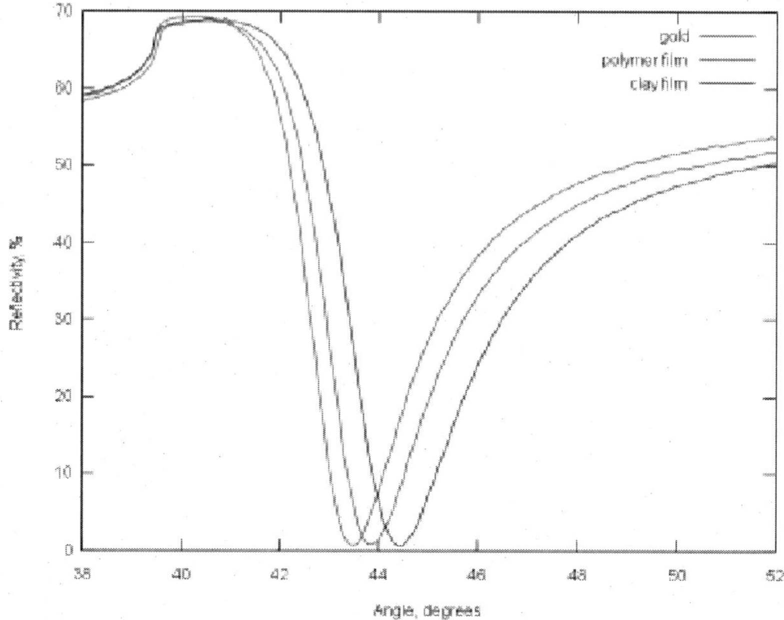

Figure 16. SPR curves measured during the adsorption of a polyelectrolyte and then a clay mineral self-assembled film onto a thin (ca. 38 nanometers) gold sensor (Source & Citation: Zeng, S., Yu, X., et al., (2012). *Sensors and Actuators B: Chemical* 176: 1128. doi:10.1016/j.snb.2012.09.073).

5.2. Binding Constant Determination

Association and Dissociation Signal

When the affinity of two ligands has to be determined, the binding constant must be determined. It is the equilibrium value for the product quotient. This value can also be found using the dynamical SPR parameters and, as in any chemical reaction, it is the association rate divided by the dissociation rate.

For this, a so-called bait ligand is immobilized on the dextran surface of the SPR crystal. Through a microflow system, a solution with the prey analyte is injected over the bait layer. As the prey analyte binds the bait ligand, an increase in SPR signal (expressed in response units, RU) is observed. After desired association time, a solution without the prey analyte (usually the buffer) is injected on the microfluidics that dissociates the bound complex between bait ligand and prey analyte. Now as the prey analyte dissociates from the bait ligand, a decrease in SPR signal (expressed in resonance units, RU) is observed. From these association ('on rate', k_a) and dissociation rates ('off rate', k_d), the equilibrium dissociation constant ('binding constant', K_D) can be calculated [43] (Figure 17).

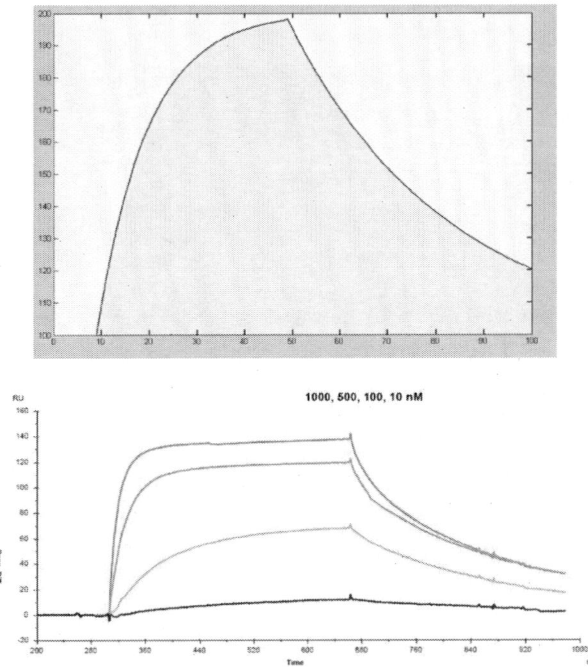

Figure 17. Example of output from Biacore (Source & Citation: Zeng, S., Yu, X., et al., (2012). *Sensors and Actuators B: Chemical* 176: 1128. doi:10.1016/ j.snb.2012.09.073).

The actual SPR signal can be explained by the electromagnetic 'coupling' of the incident light with the surface plasmon of the gold layer. This plasmon can be influenced by the layer just a few nanometer across the gold-solution interface, i.e. the bait protein and possibly the prey protein. Binding makes the reflection angle change;

$$K_D = \frac{k_d}{k_a}$$

Chapter 6

GENERAL PRINCIPLES OF SPR EXPERIMENTS

6.1. A TYPICAL EXPERIMENT

A typical SPR experiment involves several discrete tasks.

- Prepare ligand and analyte.
- Select and insert a suitable sensor chip (CM 5 or others available).
- Immobilise the ligand and a control ligand to sensor surfaces.
- Inject analyte and a control analyte over sensor surfaces and record response.
- Regenerate surfaces if necessary.
- Analyse data. While the ligand and analyte could be almost any type of molecule, they are usually both proteins. This part focuses on the analysis of protein-protein interactions.

Sensor Chips

CM 5 (Research / certificate grade)- with pre-coupled carboxymethylated dextran. Suitable for most of the application. However, research grade chip is more expensive.

Sensor chip (SA / NTA / HPA)- pre-coupled with streptavidin (SA), nitrilotriacetic acid (NTA) or flat hydrophobic surface (HPA). These are expensive and cannot be reused. HPA chip is exclusively used for immobilizing lipid monolayers.

Pioneer chip (B1, C1, F1, J1, L1) - Have low to high level of dextran matrix and carboxymethylation, gold surface or dextran modified with lipophilic compounds. Extensively used for binding bulky particles. Have reduced matrix related artefacts.

6.2. Preparation of Materials and Buffers

Air bubbles or other particles passing through the flow system frequently disrupt the SPR experiments. These can be flushed out and the experiment repeated, thus wasting time and reagents. Sometimes, the damage is irreversible necessitating the replacement of an expensive sensor chip or integrated fluidic cartridge. This can however, be minimised by following a simple rules [23].

6.3. Monitoring the Dips

The output from the photo-detector array that is used to determine the surface plasmon resonance angle (θ_{spr}) can be viewed directly as 'dips'. Protocol 1 describes how to view and interpret the dips [23].

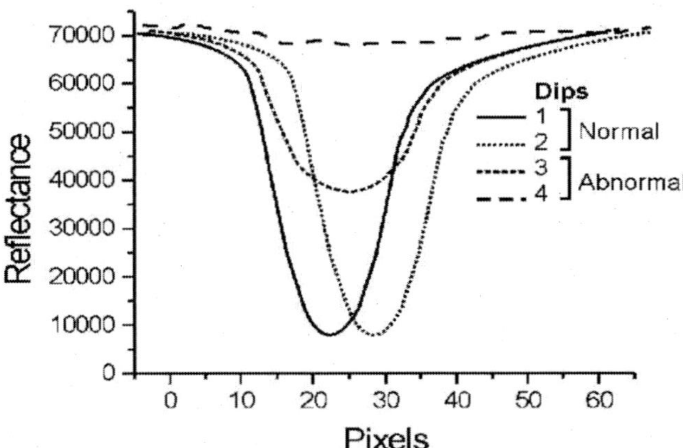

Figure 18. Analysis of dips. Normal dips shown are (1 and 2) and abnormal (3 and 4) dips.

Protocol 1. Normal and Abnormal 'Dips'

Viewing the Dips

A graph appears similar to the one in Figure 18 showing the amplitude of light reflected off the sensor surface (Reflectance) measured over a small range of angles. The angle of the minimum reflectance (θspr) is calculated by fitting a curve to all this data, thereby enabling θ_{spr} to be measured at a far greater resolution than the resolution at which the data are actually collected.

1.1. Normal Dip

Depth of dip is the most important feature of a normal dip (Figure 18). Generally it bottoms out at a Reflectance of ~10000. When the refractive index above the sensor surface increases (e.g. because of protein binding), the dip shifts to the right while maintaining its shape and depth and (Dip 2) (Figure 18).

1.2. Abnormal Dip

Two main types of abnormal dip.

1. The *shallow dip* (Dip 3). This is usually below 10000 RUs. The θ_{spr} is measured along a section of the sensor surface rather than at a single point. Refractive index heterogeneity (differences in the amount of material immobilized along the surface or small air bubbles or particles in the flow-cell) on the surface gives heterogeneous θ_{spr} values. In the former case, dip is only slightly shallow while in latter case it is very shallow (Figure 18).
2. *No dip* (Dip 4). This happens when change in the refractive index is too large and beyond the instrument dynamic range, usually due to air in the flow cell. This will shift the θ_{spr} so much that no dip is evident (Figure 18).

Slight shallowness of the dip on account of coupling large amount of proteins is acceptable. More severe abnormalities however, should not be ignored. In such cases, flow cell should be flushed with buffers or sensor surface be regenerated. If this is ineffective a new sensor surface should be used (Figure 18).

Chapter 7

LIGAND

7.1. DIRECT VERSUS INDIRECT IMMOBILIZATION

Immobilisation of proteins (the ligand) to the sensor surface without disrupting its activity is the most daunting step in the SPR experimental set up. Immobilisation by direct method involves covalent coupling, while indirect methods involves capture by a covalently coupled molecule.

Direct covalent immobilisation is advantageous in that it can be used for any protein provided that it is reasonably pure (>50%) and has a pI > 3.5. However, with three important drawbacks. Firstly, it is often difficult to regenerate the directly-coupled proteins. Secondly, binding to analyte is either decreased or completely abrogated. And thirdly because these usually have multiple copies of the functional group that mediates immobilization, proteins are coupled heterogeneously and sometimes at multiple sites

Indirect immobilisation is disadvantageous in that it can only be used for proteins that have a binding site or tag for the covalently coupled molecule. However, it has four distinct advantages, which make it the method of choice in most of the cases. Firstly, the protein need not be pure; it can be captured from a 'crude' sample, which is a big advantage for protein biologist. Secondly, proteins are seldom inactivated by indirect coupling. Thirdly, all the molecules are immobilised in a known and consistent orientation on the surface. Finally, using appropriate buffers it is possible to regenerate the surface, thereby enabling the ligand surface to be re-used, i.e. ‘regeneration’ by dissociating selectively the non-covalent ligand/analyte bond [23].

7.2. COVALENT IMMOBILISATION

7.2.1 A General Approach

Protocol 2. A General Approach to the Covalent Coupling of a Protein

1. Select the coupling chemistry.
2. Prepare the protein.
3. Optimize the pre-concentration step.
4. Couple the protein.
5. Evaluate the activity of the immobilised protein.
6. Establish conditions for regeneration.
7. Adjust the immobilization conditions.

2.1. Choice of Chemistry

All covalent coupling methods utilize free carboxymethyl groups on the sensor chip surface. They can therefore be used for any of the sensor chips that have such carboxymethyl groups. There are three main types of coupling chemistry, which utilize, respectively, amine (e.g. lysine and unblocked N-termini), thiol - ligand (cysteine /disulphide) or surface thiol (amino / carboxyl) aldehyde (carbohydrate) functional groups on glycoproteins. If the protein to be immobilised has a surface-exposed disulphide or a free cysteine, ligand-thiol coupling is preferable. Failing this, amine coupling should be tried in the first instance. If amine coupling inactivates the protein (as assessed by ligand and/or mAb binding), aldehyde coupling can be attempted, provided that the protein is glycosylated. Amine coupling can be used for most of the proteins, whereas thiol coupling requires proteins with exposed disulphides or free cysteines and aldehyde coupling is useful for glycoproteins and polysaccharides. Surface thiol coupling is not much in use. Detailed protocols are available from different instruments manual for all coupling techiques. Only amine coupling is described here in some detail [23].

2.2. Prepare the Protein

Usually a modest amount (5-10 μg) of protein is required for the protocol. Direct coupling is relatively indiscriminate, i.e. all protein in the preparation will be coupled. Therefore the major requirement is that the protein is pure and has a high level of activity. Thus, if the preparation is contaminated by other proteins or is partially active, the level of binding observed will be proportionately decreased. Because the protein needs to be diluted into the pre-concentration buffer to a final concentration of 20-50 μg/ml, the stock should be fairly concentrated (> 0.5 mg/ml). If the protein is in solutions that are strongly buffered or contain high salt concentrations, i.e. primary amine groups or sodium azide, then these must either be dialysed out or much larger dilutions (1:100) made. In the latter case the protein will need to be at higher concentrations (>2 mg/mL) [23].

2.3. Pre-concentration

The sole purpose of pre-concentration step is to concentrate the protein to very high levels (>100 mg/mL) within the dextran matrix, to drive the coupling reaction. Otherwise, higher concentrations of protein would need to be injected to get equivalent levels of coupling. An electrostatic interaction between the negatively charged carboxylated dextran matrix and positively charged protein drives the pre-concentration step. Protein is diluted into a buffer with a low ionic strength (to minimize charge screening) with a pH below its isoelectric point or pI (to give the protein net positive charge). However, at a high pH, amine coupling is most efficient because activated carboxyl groups react better with uncharged amino groups. Thus, the highest pH compatible with pre-concentration is determined empirically (**Protocol 3**). Electrostatically-bound protein should dissociate rapidly and completely when the injection of running buffer resumes, because the proteins net positive charge will decrease and electrostatic interactions will be screened by the high ionic strength of the running buffer. Incomplete dissociation suggests that the interaction was not purely electrostatic, perhaps because of the binding at low pH of a denatured form of the protein [23].

Protocol 3. Determining the Optimum pre-Concentration Conditions for a Protein

1. Protein should be diluted to the final concentration of 20-50 μg/mL (final volume 100 μL) into pH 6.0, 5.5, 5.0, 4.5 and 4.0 pre-concentration buffers.
2. Start a manual SPR run using a single flow cell at the flow-rate of 10 μL/min
3. Inject 30 μL of each sample, starting with pH 6.0 and going down.
4. In case no electrostatic interaction is observed then continue with protein samples diluted in further lower pH 3.5 and 3.0 pre-concentration buffers.
5. Use the highest pH at which greater than 10000 RU of protein binds electro-statically during the injections.

6. Check that all the bound protein dissociates after the injection. If not it suggests that not all the protein was bound electrostatically and that the protein denatures irreversibly at that pH.

2.4. Amine Coupling

The standard protocol for amine coupling involves three main steps of Activaion, Coupling, Blocking and Regeneration (Protocol 4). *Activation Step*, activation of carboxymethyl groups with N-hydroxysuccinimide (NHS), thus creating a highly reactive succinimide ester which reacts with amine and other nucleophilic groups on proteins. *Coupling Step,* to inject the protein in pre-concentration buffer, to get high protein concentrations and driving the coupling reaction. *Blocking Step,* blocks the remaining activated carboxymethyl groups by injecting very high concentrations of ethanolamine. The high concentration of ethanolamine also helps to elute any non-covalent bound material. *Regeneration Step* is optional; when protein is coupled for the first time it is advisable not to include any regeneration step. Structural integrity of the protein should be evaluated before regeneration is attempted. If it is included and the protein has poor activity it will not be clear whether covalent coupling or regeneration was responsible for disrupting the protein. Once regeneration conditions have been established these can be added on to the coupling protocol [23].

Protocol 4. Amine Coupling

Reagents

120 μL of protein at 20-50 μg/ml in suitable pre-concentration buffer (Protocol 3)

120 μL of EDC (0.4 M) mixed 1:1 with NHS (0.1 M). Make up just before coupling

120 μL of ethanolamine/HCl 1 M, pH 8.0

120 μL of regeneration solution (if regeneration conditions known)

Procedure

1. Establish pre-concentration conditions (Protocol 3)
2. Set flow-rate to 10 μL/min
3. *Activation:* inject 70 μL of EDC/NHS
4. *Coupling:* inject 70 μL of protein
5. *Blocking:* inject 70 μL of ethanolamine
6. *Regeneration:* inject 30 μL of regeneration solution (if regeneration conditions known, see Protocol 5).

7. View the 'dips' (Protocol 1) to confirm that the immobilisation is homogeneous.

2.5. Regeneration

Regeneration allows surfaces to be re-used many times, saving both time and money. Establishing ideal regeneration conditions can be a very time-consuming, and in many cases a daunting task. Thus, it may be more cost-effective to opt for imperfect or no regeneration, using new sensor surfaces instead.

Regeneration can be attempted, once a covalently immobilised protein has been shown to be active with respect to binding its natural ligand or a monoclonal antibody. A general approach to establish regeneration conditions should be used (Protocol 5) with a view to elute any non-covalently bound analyte without disrupting the activity of the ligand [23].

Protocol 5. Establishing Regeneration Conditions

Regeneration solutions

- There are four main classes: divalent cation chelator, high ionic strength, low pH, high pH.
- If the interaction is likely to be dependent on divalent cations try buffers with EDTA.
- If the monomeric interaction is weak, try high ionic strength buffers.
- Otherwise start with a low pH buffer.
- If this is without effect, try high pH buffer.

Approach

1. Covalently couple ligand to the surface.
2. Make up enough analyte for several injections.
3. Inject analyte over the immobilised ligand and measure the amount of binding.
4. Inject regeneration buffer for 3 min, and measure the amount of analyte that remains bound.
5. If the decrease is less than 30% then switch to a different class of regeneration buffer and return to step 4.
6. If the decrease is greater than 30% but less than 90%, try repeated injections. If this fails to elute greater than 90% select a stronger regeneration buffer of the same type and return to step 4.
7. When greater than 90% of bound analyte is eluted, return to step 3, using identical analyte and injection conditions. If after regeneration

binding remains greater than 90% of binding before regeneration, the regeneration conditions may be adequate.

8. If residual activity is less than 90%, select a different class of regeneration buffer. If residual binding is very low it may be necessary to return to step 1, starting with a new surface.
9. If two different types of buffer give partial elution, try using both sequentially.

2.6. Adjusting the Immobilisation Conditions

Usually the level of ligand coupling achieved is unpredictable therefore it is necessary to modify the initial protocol to achieve the desired immobilization level. It is often necessary to create several surfaces on the same chip with different levels of coupling. The best way to achieve different levels of coupling is to change the duration of the activation step, by varying the volume of NHS/EDC injected. The level of coupled ligand varies in proportion with the duration of the activation step. Thus, a two-fold reduction in activation period will usually lead to a two-fold reduction in coupling. It is possible using an option in the inject command to couple ligand simultaneously in multiple flow cells, varying only the length of activation step [23].

7.3. Non-covalent Immobilisation (Ligand Capture)

Two chief requirements for non-covalent or indirect immobilization of a ligand (henceforth called 'ligand capture') is to obtain or create a sensor surface (it should be possible to regenerate this surface so that repeated capture is possible) that can capture a ligand.

Second, the ligand needs to have a suitable binding site or modification that allows it to be captured. Second, the ligand needs to have a suitable binding site or modification that allows it to be captured [23].

7.4. Using an Existing Strategy

A number of optimized techniques are available for ligand capture, e.g. if covalently coupled molecule is anti GST (captured ligand will be any protein expressed as GST chimeras), for streptavidin (any biotinylated molecule), Ni-

NTA (any protein with oligo histidine tag), mouse anti-human Fc monoclonal (most Fc chimeras) and rabbit anti-mouse Fc polyclonal (any mouse IgG monoclonal antibody). When recombinant protein is used for SPR studies it is advisable to add a suitable domain or peptide motif so that one of these techniques can be used. The precise choice of tag will depend on whether the tag is needed for purification and what other uses are envisaged for the ligand.

- If it is envisaged that the ligand is to be used as an analyte in SPR studies it should be monovalent or, if multivalent (Fc- and GST-chimeras), the tag should be readily removable.
- Removable tag should be used if the ligand is to be used for structural studies.
- If the ligand is to be used to probe for binding partners on cells or tissues it should be multivalent, or it should be possible to make it multivalent.
- The capturing agent needs to have a high affinity and to be highly specific if the ligand is to be captured from crude mixtures (i.e. after expression without purification). For example, CD4 [**37**] and anti-human IgG_1 [36] mAbs have been shown to be suitable for this purpose.

7.5. Developing a New Strategy

Monoclonal or polyclonal Abs can tolerate harsh regeneration conditions; this coupled with widespread availability of purified monoclonal or polyclonal Abs make them suitable reagents for non-covalent immobilization. Typically there is a need to immobilize a number of related ligands, with an invariant and a variant portion. If several antibodies against the invariant portion are available, it is likely that at least one of these will be suitable for indirect coupling. A standard protocol to develop such a method is described in Protocol 6.

Protocol 6. Developing a New Antibody-Mediated Indirect Coupling Method

1. Obtain a panel of antibodies available in pure form that can bind a suitable ligand.
2. Analyse the binding properties and select a high affinity antibody with a slow k_{off}.

3. Covalently couple the antibody by amine coupling (Protocol 4) and check for activity. If inactive, try a different antibody.
4. Establish regeneration conditions (Protocol 5).
5. If suitable regeneration conditions cannot be found, try a different antibody.

7.6. Activity of Immobilised Ligand

It is important to evaluate the functional integrity of the immobilised protein. The ability to bind its natural ligand is reassuring evidence that an immobilised ligand is functionally intact. This is best achieved by using a protein, which binds to correctly-folded ligand. MAbs are particularly useful if the natural binding partner has not been identified and/or a candidate binding partner being assessed for an interaction with the ligand. Since mAbs usually bind to 'discontinuous epitopes' on a protein, they are excellent probes of protein structure. It is preferable to check the binding of several mAbs, including ones that bind to the same binding sites or, failing this, the same domain of the natural ligand. It is also important to know whether immobilised ligand is 'active' and also what proportion is active Protocol **7** [23].

Protocol 7. Quantitating Binding Levels

The ligand activity (ActL), stoichiometry (S), molecular mass of ligand (M_L) and molecular mass of analyte (M_A), and the analyte binding level at saturation (A) are related as follows:

$$ActL * S = (M_L / M_A) * A / L$$

Where, L is the ligand immobilized

S is the molar ratio of analyte to ligand in analyte/ligand complex

The product ActL*S is readily calculated, once A and L have been measured. Either ActL or S needs to be independently determined in order to calculate the other. A convenient way to measure ActL by SPR is to use Fab fragments of mAbs specific for the ligand. Intact mAbs are less useful because of the uncertainty of their binding stoichiometry [23].

7.7. Control Surfaces [23]

A control surface similar to the ligand surface should be generated, including similar levels of immobilisation. This is to quantitate non-specific binding and to record the background response. The immobilisation levels need to be similar because this affects the background response measured with analytes that have a high refractive index.

7.8. Re-using Sensor Chips

Because each sensor chip has several flow cells, it is common to have unused flow cells at the end of an experiment. In addition many covalently coupled ligands are very stable, enabling surfaces to be re-used over several days. It is therefore convenient to be able to remove and reinsert sensor chips in the instrument (Protocol 8).

Protocol 8. Reusing Sensor Chips

- Undock the sensor chip, choosing the empty flow cell option.
- Store the sensor chip in its cassette at 4°C.
- When it is to be re-used, re-insert and dock the sensor chip.
- After priming the system check the 'dips'.

Chapter 8

ANALYTE

The extent to which the analyte needs to be characterised is decided by whether, quantitative measurements are being done. This require that the analyte is very well characterised and of the highest quality. In contrast, this is less important for qualitative measurements.

8.1. PURITY, ACTIVITY AND CONCENTRATION

Concentration of the injected material should be known with great precision to determine the affinity and association rate constants. This can be achieved by using a spectrophometer to measure the optical density of a solution of pure protein, usually at 280 nm (OD_{280}). This coupled with the determination of extinction coefficient, which can be calculated directly by amino acid analysis of a sample of the protein with a known OD_{280} using online software, 'Protein Calculator'v3.3 [http://www.scripps.edu/~cdputnam/protcalc.html]. Also, what proportion of the purified protein is 'active', i.e. able to bind to the ligand can be determined by depletion experiments in which the ligand-coated sepharose beads are used to deplete the analyte from solution [43]. MAb coated sepharose beads can be used if ligand-coated beads are impractical. If all the analyte can be depleted in this way it is 100% active [23].

8.2. VALENCY

Preferably, for affinity and kinetic measurement analyte molecule should have a single binding site, i.e. monovalent. If the protein has a single binding site it is only necessary to show that it exists as a monomer in solution. This is most readily achieved by size-exclusion chromatography / Gel Filtration chromatography using Fast Pressure Liquid Chromatography (FPLC) or analytical ultracentrifugation. However, these analytical techniques will not detect the presence of very low concentrations of multivalent aggregated material [44, 45]. To ensure that this material does not contribute to the binding it is essential to purify the monomeric peak by size exclusion chromatography before SPR experiments and to analyse it before concentrating or freezing it. Only when it has been shown that concentration, storage or freezing do not affect the measured affinity and kinetic constants is it wise to deviate from this strict principle. Fortunately, the presence of multivalent aggregates is readily excluded by analysis of the binding kinetics. If the dissociation of bound analyte is monophasic (mono-exponential) multivalent binding can be ruled out. If dissociation is bi-exponential with greater than 10 fold difference in the two k_{off} values, multivalent binding is likely. Bi-exponential dissociation with smaller differences in the two k_{off} values could be due to several reasons.

8.3. REFRACTIVE INDEX EFFECT AND CONTROL ANALYTES

When an analyte is injected over a surface it is important to perform a second injection with a control analyte. This will rule out the non-specific binding and it also controls for any refractive index artefacts. This may occur when the background signal measured during the injection of an analyte sample differs between flow cells. Such an artefact will create problems for affinity measurements. It can occur whenever there is a substantial difference between surfaces, i.e. if very different levels of material are immobilised on each surface. In this case, because the immobilised material displaces volume, the volume accessible to the injected analyte sample will differ between flow-cells. If the analyte sample has a higher refractive index than the running buffer, a larger background signal will be seen from the surface with less immobilised material. This artefact is greater when (i) there are big differences in the levels of immobilised material (e.g. greater than 2000 RUs), (ii) the background signal is very large (e.g. greater than1000 RUs), and (iii) the binding response is much smaller (lesser than 10%) than the background response. These conditions are common when measuring very weak interactions, using high concentrations of analyte injected, and low molecular weight analytes, which give a very small response. A refractive index artefact can be detected by injecting a control solution with a similar refractive index to the analyte sample. If the control solution gives the same response in both flow-cells, a refractive index artefact can be excluded. Refractive index artefacts are most easily avoided by taking care to immobilise the same amount of total material on both the control and ligand sensor surfaces [23].

8.4. Low Molecular Weight Analytes

Signal to noise ratios improvements in the SPR instruments have enabled binding to be detected of analytes with M_r as low as 180 [46]. Usually, there are two major problems associated with such type of studies. One, very high densities of ligand must be immobilized in order to detect binding. The levels can be calculated using equation in Protocol 7. For an analyte of M_r ~200 that binds a ligand with an M_r of ~40000, approximately 10000 RU of active ligand needs to be immobilized to see 50 RU of analyte binding. Achieving this level of immobilization is very difficult. Also, with such small binding responses refractive index effects become significant, which can be avoided by dissolving and/or diluting the analyte in the running buffer and using a control flow cell with very similar levels of immobilization [23].

Chapter 9

QUALITITATIVE ANALYSIS; DO THEY INTERACT?

9.1. POSITIVE AND NEGATIVE CONTROLS

The main purpose of a qualitative analysis is to find out whether or not there is an interaction between a given analyte and ligand. If binding is detected it is necessary to test negative controls to exclude a false positive. These include negative ligand controls and negative analyte controls. Blocking experiments, using molecules known to block the interaction, are also useful negative controls. In contrast, if no binding is detected it becomes necessary to run positive controls to find out whether this reflects the absence of an interaction, a very low affinity, or an artefact resulting from defective ligand or analyte. The ligand can be assessed by showing that it can bind to one or more mAbs or additional analytes if they exist. The analyte can be tested by injecting it over a surface with either a known interacting ligand or mAb immobilized to the surface [23].

9.2. QUALITATIVE COMPARISONS USING A MULTIVALENT ANALYTE

The binding of a multivalent analyte can be affected by the level of immobilized ligand, since the latter will influence the valency of binding. Thus, when comparing the binding of a multivalent analyte to different ligands, it is important that these ligands are immobilized at comparable surface densities.

9.3. Quantitative Measurements

Quantitative measurements are particularly more complicated than qualitative measurements because of the quality and amount of materials required and also the difficulties associated with designing the experiments and analyzing the data. When undertaking these measurements it is important to understand the various hurdles and how these can be overcame [16,17,20]. Any quantitative analysis on the SPR requires that the analyte binds in a monovalent manner. Many binding parameters are temperature dependent so, it is important to perform key measurements at physiological temperatures (i.e. 37°C in mammals) [23].

Chapter 10

AFFINITY

10.1. CONCEPTS

There are a number of ways to represent the affinity of an interaction.

• The 'association constant' (K_A) or affinity constant is simply the ratio at equilibrium of the 'product' and 'reactant' concentrations. Thus, for the interaction A + B ↔ AB

$K_A = C_{AB} / C_A C_B$
Note that K_A has units M^{-1} (i.e. $L.mol^{-1}$)

• Many prefer to express affinity as the 'dissociation constant' or K_D, which is simply the inverse of the K_A, and therefore has the units M.

• Affinity can also be expressed as the binding energy or, more correctly, the standard state molar free energy (ΔG°). This can be calculated from the dissociation constant as follows:

$$\Delta G° = RT \ln K_D / C°$$

where T is the absolute temperature in Kelvin (298.15 K = 25 °C)
R is the Universal Gas Constant (1.987 $cal.K^{-1}.mol^{-1}$)
C^o is the standard state concentration (i.e. 1 M)

10.2. Experimental Design

The affinity constant can be measured directly by equilibrium binding analysis, or calculated from the k_{on} and k_{off}. But due to the difficulties associated with obtaining definitive kinetic data on the SPR instruments, equilibrium binding analysis is more reliable. It involves injecting a series of analyte concentrations and measuring the level of binding at equilibrium. The relationship between the binding level and analyte concentration enables the affinity constant to be calculated [19]. A basic approach to such measurements is outlined on Protocol 9.

10.3. Preliminary Steps

Equilibrium measurements are usually unaffected by mass-transport or re-binding artefacts, therefore high levels of immobilization can be used to increase the binding response. This is particularly useful when the background signal is high or the analyte very small.

Ideally, for equilibrium affinity measurements the level of active ligand on the surface should be the same for each concentration of analyte injected. This is usually straightforward when the ligand is covalently coupled (and so does not dissociate) and the analyte dissociates spontaneously within a few minutes (so that regeneration is not required). Where regeneration is required it must be shown that ligand activity is unaffected by repeated regeneration. Where captured ligands dissociate spontaneously or require regeneration, it may be difficult to maintain the level of the ligand constant. It is possible to correct for this if the level of active ligand can be accurately monitored [47].

It is important to ensure that the analyte injections reach equilibrium. While the approximate time it takes to reach equilibrium can be calculated, it is advisable to measure this directly in preliminary experiments under the same conditions (flow rate, analyte concentration, ligand density) as those to be used for the affinity measurements. Enough time must be allowed following the injection for the bound analyte to dissociate completely from the sensor surface. If dissociation is incomplete, or takes too long, it may be necessary to enhance dissociation by injecting regeneration solution.

10.4. The Experiment

Ideally the analyte concentration should be varied over four orders of magnitude, from 0.01* K_D to 100 * K_D. However, it is often only practical to vary the concentration over 2-3 orders of magnitude. This can be achieved with 10 two-fold dilutions starting at between 10 * K_D and 100 * K_D.

An assumption in these affinity measurements is that the level of active immobilized ligand remains constant. This should be checked by showing that a reference analyte binds to the same level at the beginning and end of the experiment. A second internal control is to reverse the order of injections. An efficient way of doing this is to work up from the lowest concentration of analyte, give one injection at the highest concentration, and then work back down to the lowest concentration. The same affinity should be obtained irrespective of the order of injections [23].

Chapter 11

DATA ANALYSIS

11.1. THEORETICAL CONCEPTS

To derive affinity constant from the data, simplest (Langmuir) model (A+L↔AL) is applicable in the vast majority of cases. It assumes that the analyte (A) is both monovalent and homogenous, so is the ligand (L), also that all binding events are independent. Under these conditions data should conform to the Langmuir binding isotherm,

$$\text{Bound} = C^A \text{ Max} / C^A + K_D$$

where "Bound" is measured in RUs and "Max" is the maximum response (RUs).

C^A is the concentration of injected analyte and K_D is in the same units as C^A (normally M)

The K_D and Max values are best obtained by non-linear curve fitting of the equation to the data using a suitable computer software such as Origin (MicroCal) or Sigmaplot.

A Scatchard plot of the same data obtained by plotting Bound/C^A against bound, is useful for visualizing the extent to which the data conform to the Langmuir model. A linear Scatchard plot is consistent with the model. Scatchard plots alone should not be used to estimate affinity constants since they give undue weightage on the data obtained with the lowest concentrations of analyte, which is generally the least reliable.

Non-linear Scatchard plots indicate that the data do not fit the Langmuir model. Before considering models that are more complex it is important to exclude trivial explanations [23].

11.2. Controls

Several artefacts like, effect of ligand immobilization on the binding, an error estimating the active concentration of analyte and an incorrect assumption that the analyte is monovalent can result in erroneous affinity constants. The most rigorous control is to confirm the affinity constant in the reverse or 'upside down' orientation since this excludes all three artefacts. If this is not possible the experiment should be repeated with the ligand immobilized in a different way, which addresses the possible effects of immobilization on binding. The affinity should be confirmed with two independently-produced batches of protein and with different recombinant forms of the same proteins [23].

11.3. Non-linear Scatchard Plots

A non-linear Scatchard plot does not conform to the Langmuir model. Other binding models can be invoked to explain non-linear Scatchard plots. Distinguishing between these models can be very difficult. A 'concave up' Scatchard plot is the most common deviation from linearity. This may be a consequence of heterogeneous ligand, multivalent analyte, or (rarely) negative cooperativity between binding sites. A trivial cause of analyte heterogeneity is the presence of multivalent analyte. Ligand heterogeneity may be a consequence of immobilization. A 'concave down' Scatchard plot indicates either positive cooperativity between binding sites or self-association of the analyte in solution or on the sensor surface.

The control experiments will help to eliminate some trivial explanations. If analyte has a multivalent component, the non-linear Scatchard plot will not be evident in the reverse orientation. The shape of the Scatchard plot will also depend on the surface density of the ligand. If the ligand immobilization is responsible for heterogeneity, this should be eliminated in the reverse orientation if the ligand is immobilized indirectly [48,49].

Protocol 9. Affinity Measurements

9.1. Preliminary Steps

1. Immobilize the ligand and a control. High levels of immobilization are acceptable.

2. Ensure that the analyte is monomeric and binds monovalently. Determine accurately the concentration of the analyte and the proportion that is active.
3. Determine the time it takes to reach equilibrium and the time it takes for the bound analyte to dissociate completely from the sensor surface. While this should be done empirically, under the conditions to be used for the equilibrium measurements, approximate times can be calculated from the k_{off}.
4. If necessary, establish regeneration conditions (Protocol 5).
5. Obtain a rough estimate of the K_D by injecting a series of five-fold dilutions.

9.2. Measurements

1. Prepare a dilution series of analyte starting at 10-100 times the K_D with at least 9 two-fold dilutions thereof. This should be enough for two injections at each concentration except the highest concentration, where only enough for one injection is required. A minimum of 17 μL of sample is required per injection.
2. Make up separate control sample of analyte (at concentration ~ K_D) enough for 2 injections.
3. Set the flow rate. To conserve sample this can be as slow as 1 $\mu L min^{-1}$.
4. Inject the control sample.
5. Inject the dilution series starting from low and moving up to the highest concentration (low-to-high) and then moving back down to the low concentrations (high-to-low). Inject the highest concentration only once. It is important to inject for a period sufficiently long to reach equilibrium. Either enough time must be allowed for spontaneous dissociation or the analyte must be eluted with regeneration buffer.
6. Repeat the injection of the control sample.
7. For all injections, measure the equilibrium response levels in the ligand and control flow cells. The difference between these two is the amount of binding at each concentration [23].

9.3. Data Analysis

1. Plot the binding versus the concentration for both the low-to-high and the high-to-low series.

2. Fit the Langmuir (1:1) binding isotherm to the data by non-linear curve fitting. Use this to determine the K_D and maximal level of binding.
3. Do a Scatchard plot and check if it is linear. The points on the plot where binding is less than 5% of maximum are highly inaccurate and should be ignored.
4. If the Scatchard plot is not linear do further experiments to establish cause.

Controls

1. The affinity constant should be confirmed in the reverse orientation.
2. Should this be impossible (e.g. because the ligand is multivalent) the affinity constant should be confirmed with the ligand immobilized by a different mechanism, preferably by ligand capture.
3. Use at least two independently produced batches of protein.
4. Use different recombinant forms of the same proteins.

Both the time taken to reach equilibrium and the time it takes for the bound analyte to dissociate are governed primarily by the k_{off}. For the simple 1:1 model binding will reach 99% of the equilibrium level within $4.6/k_{off}$ seconds. Similarly, it will take $4.6/k_{off}$ seconds for 99% of the analyte to dissociate. Thus, for k_{off} ~0.02 s^{-1}, equilibrium will be reached within ~ 230 s and the bound analyte will take ~230 s to dissociate.

If the Quickinject command is used as little as 15 μL of analyte sample is used.

Chapter 12

KINETICS

12.1. CONCEPTS

Association phase is the period during which analyte is being injected whereas *Dissociation phase* is the period following the end of the injection. During the association phase there is simultaneous association and dissociation. Equilibrium is reached when the association rate equals the dissociation rate. Under ideal experimental conditions only dissociation should take place during the dissociation phase, however, in reality some re-binding (see below) often occurs [23].

Factors affecting the association rate -
Concentration of analyte near the ligand (C^A),
Concentration of ligand (C^L),
Association rate constant (k_{on}).

Limitation due to Mass Transport

Because of the high surface density of ligand on the sensor surface, the rate at which analyte binds ligand can exceed the rate at which it is delivered to the surface called as mass transport. In this situation binding is said to be mass transport limited. Analysis of association rate under mass transport limited conditions will yield an apparent k_{on} that is slower than the true k_{on}. It is difficult to determine the k_{on} under these circumstances, and so experimental conditions must be sought in which mass-transport is not limiting.

Analyte is transported to the surface by both convection and diffusion. Convection transport can be increased simply by increasing the flow rate. However, mass transport can still be limiting even at the maximal flow-rates permissible [22, 43] because there is an unstirred 'diffusion' layer near the sensor surface through which transport is solely by diffusion [17]. In this case mass transport limits can only be avoided by decreasing the surface density of immobilized ligand.

Mass transport limitations, which lead to an underestimation of the intrinsic kinetics, are aggravated by low flow rates, high levels of immobilized ligand, and high intrinsic association rate constants. They can be reduced by increasing the flow rate and, most importantly, lowering the level of immobilized ligand.

Factors affecting the Dissociation rate are -

Surface density of bound analyte, Dissociation rate constant (k_{off}), and the *Rebinding,* i.e. the extent to which dissociated analyte rebinds to ligand before leaving the sensor surface. The latter is also a consequence of mass transport deficiency but here it is transport away from the surface that is limiting. Rebinding will occur even after increasing the convection transport by increasing the flow rate because diffusion out of the unstirred layer is little affected by convection transport. Re-binding is most easily avoided by decreasing the level of ligand immobilized on the surface. An alternative method is to inject during the dissociation phase a competing molecule that can rapidly bind to free analyte or ligand and block re-binding [50]. If it binds ligand the competing molecule needs to be small so that it does not influence the SPR signal. Finally, when the ligand is saturated the initial part of the dissociation phase will not be affected by re-binding, since no free ligand is available for re-binding. However such selective analysis of a part of the dissociation phases should be avoided; it provides no indication as to whether the data conforms to any particular binding model and can give highly misleading results.

12.2. EXPERIMENTAL DESIGN

Mass transport may limit binding so it is essential to use the lowest density of ligand that gives an adequate level of analyte binding. Depending on the background response, 100 RU of binding should be adequate. To determine whether binding is limited by mass transport the kinetics should be measured in several flow cells with different levels of immobilized ligand. The immobilization level should vary at least two-fold.

Protocol 10. Kinetic Measurements

Preliminary Steps

1. Immobilize the ligand in three flow cells at different levels (e.g. 500, 1000, and 2000 RU).
2. Immobilize a control ligand in the remaining flow cell at a level midway between the ranges of immobilization levels in the other three flow cells.
3. Determine accurately the concentration of the analyte. Ensure that the analyte is monomeric and monovalent.
4. Determine the time it takes to reach equilibrium and for the bound analyte to dissociate completely from the sensor surface. In less than 4-5 seconds (fast) it will be necessary to collect at the maximal rate possible (10 Hz). This is only possible if data is collected from one flow cell at a time. Because the sample needs to be injected once for each flow cell studied, more samples are needed.
5. Establish regeneration conditions if necessary.

Measurements

1. Prepare a two-fold dilution series of the analyte ranging from concentrations of $8*K_D$ to approximately $0.25*K_D$. Take care to prepare enough samples for the special kinetic injection command (KINJECT), which utilizes more material. If kinetics is fast and a high data collection rate is needed, enough analyte needs to be prepared for separate injections in each flow-cell.
2. Set the flow rate to 40-100 μL/min to maximize analyte mass transport. The duration of the injection is not critical since binding does not need to reach equilibrium. However, equilibrium should be approached at the higher analyte concentrations.
3. Inject the dilution series in any particular order. It is usual to start from lower concentrations.

Data Analysis

1. Use the software compatible with the SPR instruments, e.g. BIA evaluation software supplied by BIAcore.
2. Subtract the response in the control flow cell from the responses in each of the ligand flow cells.

3. Group and analyse together the binding curves obtained with each dilution series, one flow cell at a time (with control responses subtracted).
4. If equilibrium is reached within 1 second the association phase will not produce useful data. In this case only the dissociation phase should be analysed.
5. Attempt a global fit of the simple 1:1 binding model to the entire series of curves. Include in the fit as much of the association and dissociation phase as possible.
6. Repeat the analysis with data obtained at the other levels of ligand immobilisation. In order to prove that binding is not limited by mass transport it is necessary to show that the same rate constants are obtained at two different ligand immobilization levels. If this is not possible the measured rate constants should be considered to be lower limits of the true rate constants [43, 45].
7. If poor fits are obtained using the simple 1:1 binding model, the binding is considered complex. After excluding trivial explanations, like mass transport limitations, drifting baseline, bulky refractive index artefacts, rebinding, heterogenous immobilization, analyte is multimeric or analyte is monomeric but contaminated with multimeric aggregates, heterogenous analyte, heterogenous ligands and rarely two state binding where, ligand-analyte complex inter converts into other forms with different kinetic properties, an attempt should be made to establish and readdress the cause.

Controls

1. Confirm the kinetics constants in the reverse orientation with the ligand immobilised in a completely different manner.
2. Results should always be confirmed using separate batches of recombinant protein or different recombinant forms of the same protein.

Kinetic analysis require more analyte because the experiments are performed at a high flow-rate, and the KINJECT command wastes more material, and separate injections may be required for each flow-cell.

12.3. Data Analysis

Analysis of kinetic data is best performed using the software compatible with the SPR instruments, e.g. BIAevaluation software provided by BIAcore [21] and CLAMP (available at http://www.hci.utah.edu/cores/biacore/) [51]. While a complete discussion of kinetic theory is beyond the scope of this book, a basic approach to kinetic analysis is provided instead. At first subtract the background responses (obtained in the control flow-cells), fit the data to simple 1:1 Langmuir binding model. For any particular sensorgram include as much data as possible. This should include the entire association and dissociation phases, omitting the 'noisy' few seconds at the beginning and end of the analyte injection. Noise in the dissociation phase can be reduced by using the KINJECT command. Always fit both the association and dissociation phases simultaneously rather than separately. However the association phase cannot be analysed if equilibrium is attained within 2-4 s, which is usually the case if the k_{off} is > 1 s^{-1}. In contrast, the dissociation phase can be analysed even if the k_{off} is >1 s^{-1} [34, 47]. A rigorous test of the binding model is to fit it simultaneously to multiple binding curves obtained with different analyte concentrations. This global fitting [20, 21] establishes whether a single 'global' k_{on} and k_{off} provide a good fit to all the data. An important internal test of the validity of the kinetic constants is to determine whether the calculated K_D (K_{Dcalc}= k_{off}/k_{on}) is equal to the K_D determined by equilibrium analysis.

Binding kinetics are considered complex when a poor fit is obtained to the data using the simple 1:1 binding model. Simple 1:1 binding model predicts both the association and dissociation phases to be mono-exponential, i.e. described by an equation with a single exponential term. Whereas all complex binding models generate equations with two or more exponential terms, so when a poor fit is obtained with simple 1:1 binding model, same will get good fit with equations using two exponential terms (bi-exponential fit). Most likely explanation for this is experimental artefact; initial efforts should be directed at excluding trivial causes. Only when trivial explanations have been excluded should any effort be expended on trying to establish what complex binding model explains the kinetics. This is a difficult and often impossible task [20, 21]. It is usually impossible to distinguish between different models by curve fitting alone. Instead further experiments need to be performed.

12.4. Controls

The same controls should be performed as for the affinity measurements (Protocol 10).

12.5. Stoichiometry

The binding stoichiometry can be determined if the molecular mass of ligand and analyte are known and the activity of the ligand is known. The basic approach is to immobilise a defined amount of ligand and then saturate this with analyte. The stoichiometry can then be calculated according to Protocol 7. Because it is very difficult to saturate with analyte, the maximum level of analyte binding is best obtained by doing a standard equilibrium affinity determination. A fit of the simple 1:1 binding model to this data yields the maximum level of analyte binding as well as a K_D. The key problem is establishing the activity of the immobilised ligand. If the ligand has 100% activity in solution and is immobilised by ligand capture, it is reasonable to assume that it is all active. Activity levels can also be determined using a Fab fragment of a mAb specific for the ligand. Finally, the stoichiometry should be identical when measured in the reverse orientation.

Chapter 13

THERMODYNAMICS

13.1. CONCEPTS

The binding energy or affinity can be calculated from changes in enthalpy (heat absorbed or ΔH) and entropy (increased disorder or ΔS).

$$\Delta G = \Delta H - T*\Delta S$$

While ΔS cannot be measured, ΔH (or the heat absorbed upon binding) can be measured directly, by micro calorimetry (ΔH_{cal}), or indirectly, by van't Hoff analysis (ΔH_{vH}). If it is assumed that ΔH and $\Delta S°$ are temperature-independent, the linear form of the van't Hoff equation can be used.

$$\text{In } K_D / C° = \Delta H_{vH} /RT - \Delta S°/R$$

where C^o is the standard state concentration (1 M) and $\Delta S°$ is the change in entropy in the standard state. K_D is measured over a range of temperatures and $\ln(K_D/C^o)$ plotted against 1/T. If linear, the slope of this plot equals $\Delta H_{vH}/R$. A distinct disadvantage of this approach is that ΔH varies with temperature for protein/ligand interactions and so the plot is not linear. Consequently, K_D needs to be measured over a small range around the temperature of interest, and the slope determined within this range. This is technically difficult and likely to be inaccurate. A more rigorous approach is to measure the affinity (ΔG^o) over a wider range of temperatures and then fit an integrated (non-linear) form of the van't Hoff equation to the data [52].

$$\Delta G° = \Delta H_{To} - T\Delta S_{To} + \Delta C_p (T-T_o) + T\Delta C_p \text{In} (T/T_o)$$

where T is the temperature in Kelvin (K)

To is an arbitrary reference temperature (e.g. 298.15 K)

ΔH_{To} is the enthalpy change upon binding at T_o (kcal.mol^{-1})

ΔS^o_{To} is the standard state entropy change upon binding at T_o (kcal.mol^{-1})

and ΔC_p is the specific heat capacity (kcal.mol^{-1}.K^{-1}), and is assumed to be temperature- independent.

The ΔC_p measures ΔH (and ΔS) as function of temperature. It is almost invariably negative for protein/protein interactions [53], indicating that enthalpic effects become more favourable and entropic effects less favourable as temperature increases. This negative heat capacity is believed to be the result of the disruption at high temperatures of the ordered 'shell' of water that forms over the non-polar surfaces of a macromolecule. Consequently, the favourable entropic effect of displacing the shell upon binding is reduced. And because fewer solvent bonds are disrupted at the higher temperature the net enthalpy change becomes more favourable. ΔC_p is a useful measure of the extent of non-polar surface that is buried upon binding [54]. However determining ΔC_p by van't Hoff analysis is likely to be inaccurate. A second drawback of van't Hoff analysis is that changes in temperature may also affect the interactions between the proteins and the solution components, including water [55]. If these equilibria are coupled to the protein/protein interaction they will contribute to the ΔH_{vH}, which will therefore differ from the ΔH determined by calorimetry. Because of these drawbacks, it is advisable to confirm ΔH and ΔC_p determinations by calorimetry. Unfortunately even recently developed micro calorimeters require about one hundred fold more protein than the SPR instruments. Thus, the SPR based instruments may be the only means of obtaining enthalpy and heat capacity data when limited amounts of material are available.

13.2. Activation Energy of Association and Disassociation

The k_{on} and k_{off} generally increase with temperature. The extent of this increase is a measure of the amount of thermal energy required for binding [activation energy of association (E_a^{ass})] or dissociation [dissociation (E_a^{diss})]. E_a can be determined using the Arrhenius equation. Assuming that E_a is constant over the temperature range examined, then

$$\ln k = \ln A - E_a / R^*T$$

where k is the relevant rate constant (e.g. k_{on} and k_{off}), R is the gas constant, and A is a constant known as the pre-exponential factor. E_a is determined from the slope of a plot of In k versus 1/T. E_a^{ass} and E_a^{diss} can be considered activation enthalpies, the reaction enthalpy can be calculated from the relationship

$$\Delta H = E_a^{ass} - E_a^{diss}$$

An unusually high E_a value indicates that binding and/or dissociation require the high potential energy barriers, suggesting that conformational rearrangements are required [23].

Chapter 14

ADVANCES IN SPR TECHNIQUES IN LAST ONE DECADE (2005-2014)

14.1. INTRODUCTION

In 1982, Nylander and Liedeberg demonstrated the use of SPR for biosensing. Since then, SPR is extensively used as an optical biosensor accounting for several thousand research papers encompassing the fields of academics, industry and technology [56].

SPR is real-time, label-free measurements of binding kinetics and affinity. This has distinct advantage over radioactive or fluorescent labeling methods, in terms of 1) ligand-analyte binding kinetics, that can be probed without the costly and time-consuming labeling process that may interfere with molecular binding interactions; 2) binding rates and affinity can be measured directly and 3) low affinity interactions in high protein concentrations can be characterized with less reagent consumption than other equilibrium measurement techniques; 4) Label-free detection of molecular interactions presents an attractive alternative.

The success of SPR biosensors was indicated by the growing number of commercially available instruments. Since Biacore AB (originally Pharmacia Biosensor AB) launched the first commercial SPR biosensor on market in 1990, there have been many more competing SPR instruments including IASys (Affinity Sensors), SPR-670 (Nippon Laser Electronics), IBIS (IBIS Technology BV), TISPR (Texas Instruments), etc. [57].

In this section, plasmon resonance techniques such as SPR, SPR-imaging, SPR-MS (mass spectrometry), PWR, trends in protein array technology, a potential use of SPR biosensors in proteome and optics based research and a note on covalent and non-covalent immobilization methods for the attachment of the molecules to the sensor surface will be reviewed in terms of their fundamentals, and in the latest applications with emphasis in the studies performed with membrane proteins. This section will also throw light on SPR sensors (from typical Kretschmann prism configurations to fiber sensor schemes) with micro- or nano-structures for local light field enhancement, extraordinary optical transmission, interference of surface plasmon waves, plasmonic cavities, etc. Further, advantages and disadvantages of each methodology are provided along with some of the latest accomplishments with emphasis on the area of GPCRs.

14.2. Protein Microarray/ Proteome

With the completion of the human genome project and the advent of rapid, cost-effective nucleic acid sequencing, scientific and technological interest has shifted from analysis of genes to analysis of the corresponding proteins, and, critically, their networks of interactions with other proteins [58, 59]. To achieve the ambitious goal of proteome-scale label-free kinetic assays, a new class of SPR instrument is needed that combines: 1) high imaging resolution to collect kinetics data from individual spots on high-density microarrays; 2) massively parallel multiplexing capability and a large field-of-view; and 3) a simple and robust optical design.

14.3. Nanoplasmonics/Nanohole Arrays

Nano-plasmonics has heralded a new generation of optical biosensors. Among these, metallic films perforated with nano-holes can exhibit plasmon-enhanced extraordinary optical transmission (EOT). Nano-hole-based SPR instruments have potential for quantitative antibody screening and as a general-purpose platform for integrating SPR sensors with other bio-analytical tools. In contrast to SPR set up in Biacore™, where a flat gold film atop a prism is illuminated at a steep angle in a total internal reflection mode that complicates optical design, assembly, and alignment, nano-holes in gold films can directly launch SPR from normally incident light using a simple illumination setup of a standard microscope [60]. Besides acting as a source of SP waves, nano-holes can concurrently function as nano-wells to confine supported lipid bilayer membranes, [61] a scaffold for pore-spanning lipid membranes, [62] nano-fluidic channels to transport biomolecules, [63] or as electrodes for electrochemical sensing. This instrument can quantitate binding kinetics of small antibody fragments (25 kDa) at low concentrations (1 nM) and measure kinetic parameters with dissociation constants ranging from 200 pM to 40 nM, dissociation rate constants ranging from 10^{-2} to 10^{-4} sec^{-1} [64].

Huang et al. (2007) showed that *rolling circle amplification (RCA)* and nanogold-modified tags can enhance the signals of analyte binding thereby increasing the sensitivity of the technique. *Fano type* resonant nano-structures were utilized to design chip based biosensors for label free detection [65].

14.4. Chemical Microarrays

Chemical microarrays, integrated with SPR imaging, can simultaneously generate affinity data for protein targets with up to 9,216 immobilized fragments per array. This approach has proven to be suitable for screening fragment libraries of up to 110,000 compounds in a high throughput fashion. This approach has led to successful identification of selective low molecular weight inhibitors for pharmacologically relevant drug targets through the SPR screening of fragment libraries [66].

14.5. Membrane Bound Proteins / SPR Nano-pore Arrays

Metallic nano-pore array supports free-standing lipid bilayers on a gold film. In this format, membrane proteins would be presented in a lipid bilayer that mimics the natural biological membrane to allow functional studies and label-free kinetic measurements. To interface with existing SPR instrumentation, membrane proteins can be immobilized as detergent "solubilized" protein, deposited in supported lipid bilayers or trapped in vesicles which are subsequently captured [67]. This is particularly important as about 30% of genes in the human genome are predicted to encode for membrane proteins, these molecules remain poorly characterized, largely due to difficulties in purifying protein for analysis.

14.6. SPR Imaging + Nanohole Arrays

Lee et al. (2012) [68] has demonstrated a SPR imaging spectroscopy instrument capable of extracting binding kinetics and affinities from 50 parallel microfluidic channels simultaneously. The ability of metallic nano-hole arrays to give rise to the extraordinary optical transmission (EOT) effect have been widely investigated for SPR sensing by combining nano-hole-based SPR technologies with imaging spectroscopy [69,70]. In another study Yu et al. prepared metallic corrugated structures for use as highly sensitive plasmonic sensors. Using nano-imprint-in-metal method, fabrication of the metallic corrugated structure was achieved in a single step. Compared to the other SPR sensors, corrugated Au films exhibited high sensitivity (ca. 800 nm/RIU) [71].

Ordered nano-hole arrays are typically fabricated using focused ion beam (FIB), e.g., [72], or electron beam lithography (EBL) [73]. However, even though these techniques have very high resolution, the manufacture of arrays requires long processing times and, therefore, they are not suitable for mass-production due to their low throughput and the associated high cost.

Blanchard-Dionne et al. had shown another SPR sensor based on a *rectangular nano-hole array* in a metal film. This SPR setup uses balanced intensity detection between two orthogonal polarizations of a He-Ne laser beam. This allows for sensitivity improvement, noise reduction and rejection of any uncorrelated variation in the intensity signal. A bulk sensitivity resolution of 6.4×10^{-6} RIU was also demonstrated. This technique has potential applications in portable nano-plasmonic multisensing and imaging [74].

A nano-hole based low-noise SPR was built around a standard microscope and a portable fiber-optic spectrometer utilizing extraordinary optical transmission (EOT). The said metallic nano-hole array platform quantified a broad range of antibody ligand binding kinetics with equilibrium dissociation constants ranging from 200 pM to 40 nM and refractive index resolution 3.1×10^{-6} without on-chip cooling. This was lowest among SPR sensors based on EOT [69,75].

Nano-holes in gold films can be directly used to perform SPR from normally incident light using a simple illumination setup such as a standard microscope. Nano-holes can concurrently function as nano-wells on supported lipid bilayer membranes, a scaffold for pore-spanning lipid membranes, nano-fluidic channels to transport biomolecules and as electrodes for electrochemical sensing [59, 60].

Nakamoto et al. [76 2012] constructed a device with a *circular nano-hole array* structure fabricated with a nano-imprint technique using a UV curable polymer followed by thin gold film deposition by electron beam. To check the efficacy of the device, immunochemical reaction was performed to detect TNF-α on the plasmonic array that gave the detection limit of 21 ng mL^{-1}.

Hexagonal nano-hole arrays can be used to characterize the strong influence of grating coupling on electromagnetic field distribution, frequency and degeneracy of plasmon bands [77].

Perdiguero et al. have shown the fabrication and characterization of *optical gold nano-hole array* sensors. The light transmission properties of the arrays have been tested in a designed optical setup and chemical plasmonic sensing observed using solutions with different known refractive indexes. From these measurements a sensitivity of 126 nm/RIU was obtained [78].

14.7. SPR-MS

The combination of SPR based technology with mass spectrometry (MS) has created a unique analytical tool for functional proteomics investigations. SPR-MS can be used as a unique multiplexed detection technology capable of both quantitative and qualitative protein analysis. SPR is used for protein quantification and interaction analysis, and MS is used to elucidate the structural features of the proteins. Since SPR detection is non-destructive, proteins retrieved on the SPR sensing surface can be further analyzed via MS, either directly from the sensor chip surface, or separately, following elution and micro-recovery after tryptic digestion of the recovered proteins. The SPR-MS combination enables identification of: (1) non-specific binding of other biomolecules to the surface-immobilized biomolecules (or to the non-derivatized sensor surface itself); (2) bound protein fragments; (3) protein variants (caused by post-translational modifications and point mutations); and (4) proteins complexed with other molecules. The SPR-MS arrays are robust, highly reproducible, and are capable of high-throughput analysis. An approach compatible with SPR arrays is, *on-chip MALDI-TOF MS*, from arrayed spots on SPR surface [79]. Another promising aspect of SPR-MS is *Direct on-chip MALDI-MS* detection using a SPRi-sensor biochip in a microarray format [80].

14.8. SPR Based on Nano Gratings

SPR sensor platform based on Nano gratings does not employ Kretschmann configuration or prism coupling thereby removing any coupling angle requirements as in the conventional Kretschmann based SPR sensors. It directly couples the normally incident radiation into narrow groove metallic nano-gratings that is very sensitive to localized changes in refractive index in the vicinity of the metallic surface of the nano-gratings. These one dimensional structures are very easy to fabricate and are not sensitive to temperature-dependent refractive index variations. [81].

14.9. SPR Microscopes and SPRi

SPR microscopes are capable of high-throughput kinetic studies of the binding of small (approximately 200 Da) ligands onto large protein microarrays. This can simultaneously monitor binding kinetics on >1300 spots in a protein microarray with a detection limit of approximately 0.3 ng/cm^2, or <50 fg per spot (<1 million protein molecules) with a time resolution of 1s, and spot-to-spot reproducibility. The protocol is label free and uses less biomolecules than standard SPR sensing and reveals the absolute bound amount and binding stoichiometry [82, 83].

SPR imager (SPRi) is used to measure whole cells label-free using the specific binding of cell surface antigens expressed on the surface of cancer cells and specific ligands deposited on sensor chips on IBIS MX96. SPR curves show a response that depends on sedimentation of cells and a specific binding response on top of the initial response, the magnitude of which is dependent on the ligand density and the cell type used. Comparison of SPRi with flow cytometry showed similar Epithelia cell adhesion molecule (EpCAM) expression on MCF7, SKBR3 and HS578T cells. [84].

14.10. Surface-enhanced Raman Spectroscopy (SERS)

SERS has the ability to control the size, shape, and material of a surface. Because excitation of the LSPR of a nano-structured surface or nano-particle relies on SERS, the ability to control surface characteristics has made SERS an interesting surface phenomenon for its use as an analytical tool. This technique uses nano-sphere lithography for the fabrication of highly reproducible and robust SERS substrates. SERS has potential application for the detection of chemical warfare agents and several biological molecules [85].

14.11. Signal Locked SPR Technique

This novel SPR technique measures the biomolecular interaction over a narrow frequency range under periodic excitation instead as a function of time. The measured response is, thus, locked to a very specific narrow band signal and has a very high degree of rejection to the uncorrelated spurious signals [86].

14.12. Wavelength Interrogation Surface Plasmon Resonance (SPR) Spectroscopy

Wavelength interrogation SPR spectroscopy uses a dove prism and integrates optical design with high-resolution refractive index monitoring and biosensing. This optical configuration uses a single axis optical path between each optical component with incorporation of fluidics for efficient sample delivery. A BK7 dove prism inverts an optical image with a total internal reflection angle of 72.8 degrees, an angle active in SPR. Hence, a unique system can accomplish SPR biosensing using wavelength interrogation and also perform SPR imaging [87].

14.13. Phase-sensitive Surface Plasmon Resonance (SPR) Biosensor

Wide dynamic range phase-sensitive SPR biosensor is based on *differential spectral interferometry*. In contrast to the conventional spectroscopic approach where only the SPR dip is monitored, this platform acquires the spectral phase information of the entire electromagnetic field that undergoes SPR transformation. Very high sensitivity achieved, is maintained continuously across the spectral domain in response to refractive index changes with a detection limit of 2.2×10^{-7} in terms of refractive index unit (RIU) using standard single-layer gold surface. Using this configuration, detection sensitivity of 0.5 ng $ml^{-1}$was estimated for BSA antibody-antigen binding experiments [88].

14.14. Combined Surface Plasmon Resonance and Electrochemical Impedance Spectroscopy (SP-EIS) Biosensing

SP-EIS is a promising technology for SPR mono- and multi-sensing applications in portable and stationary platforms. Combined surface plasmon resonance and electrochemical impedance spectroscopy (SP-EIS) biosensing system uses a sensor chip consisting of micrometer sized interdigitated metal electrodes (IDEs). This acts as a surface plasmon supporting structure. The IDEs exhibit the properties of an optical dispersive grating reflecting white light into the diffraction orders. Direct surface plasmon spectral analysis is possible by using the first order of diffraction and has been used as a novel principle for a simplified optical set-up [89].

14.15. Multiplex Surface Plasmon Resonance (SPR)

This platform allows measurement of replicate interactions with a single, parallel set of analyte injections, whereas repeated regeneration of the scaffold (anti-insulin receptor monoclonal antibody 83-7) between measurements causes variable loss of antibody activity. Fit to kinetic data was improved on including a conformational change factor in which the high-affinity site was converted to the low-affinity site. The new SPR assay enables insulin-eIR-A interactions to be followed in real time and has potential application to study the effects of humoral factors on the interaction, without the need for insulin labeling [90].

14.16. Coupling Microchip CE (MCE) to Surface Plasmon Resonance (SPR)

MCE is a microfluidic technique that utilizes micro fabrication to connect interacting fluid reservoirs. The goal was to develop a simple means to couple microchip CE (MCE) to surface plasmon resonance (SPR). Its advantages include rapid analysis (typically seconds), easy integration of multiple analytical steps and parallel operation. This technique can be successfully applied in the separation of two species. The method offers the advantages of a near zero connection dead volume, electrical shielding from the separation voltage and minimization of the mass transfer effect [91, 92].

14.17. SPR Based Natural Glycan Microarray

SPR based natural glycan microarray is suited for screening of interactions between glycans and carbohydrate-binding proteins (CBPs). Utility of the array was demonstrated in the simultaneous detection of glycan-specific serum antibodies, the anti-glycan antibody profiles from sera of *S. mansoni* infected individuals as well as from non-endemic uninfected controls. The SPR screening was found to be sensitive for differences between infection sera and control sera, and revealed antibody titers and antibody classes (IgG or IgM). All SPR analyses were performed with a single SPR array chip that was regenerated and blocked before the application of serum sample. Results indicate that SPR-based glycan (natural or synthetic) can be used for determining serum antibody profiles as possible markers for the infection status of an individual [93].

14.18. Nanodiscs

Nano-discs are small, flat model membranes that mimics native environment for reconstitution of integral membrane proteins. Incorporating membrane proteins into nano-discs lead to water soluble proteolipid particles. This makes the membrane proteins amenable to a number of bioanalytical techniques originally developed for soluble proteins. Results of single cycle SPR showed that nano-disc incorporated membrane protein has almost identical affinity toward the antibody as did the soluble decahistidine-tagged ubiquitin studied in a comparative experiment. [94].

14.19. SPR with Graphene Chip

This was the first ever study to use graphene in the SPR sensor in place of gold/silver. Graphene in the SPR sensor is expected to broaden the range of analyte to bio-aerosols based on the superior electromagnetic properties of graphene. This sensor is based on the resonance condition of the surface plasmon wave (SPW), dielectric constants of each metal film, detected material in gas or aqueous phase and fiber optic coated with multi-layered graphene. The multi-layered graphene film synthesized by chemical vapor deposition (CVD) was transferred on the sensing region of an optical fiber. The graphene coated SPR sensor is used to analyze the interaction between structured DNA biotin and Streptavidin. Results show excellent detection sensitivity of the interaction between structured DNA and Streptavidin [95].

14.20. Matrix-free carboxylated Surface Plasmon Resonance (SPR) Sensor Chip

A very high sensing matrix-free carboxylated SPR sensor chip was designed by functionalizing a bare gold thin film with a self-assembled monolayer of 16-mercaptohexadecanoic acid (SAM-MHDA chip). SAM was characterized by infrared reflection absorption spectroscopy, X-ray photoemission spectroscopy, atomic force microscopy, X-ray reflectivity-diffraction, and SPR experiments with bovine serum albumin. SPR signal were compared on this chip made of a dense monolayer of carboxylic acid groups with commercially available carboxylated sensor chips built on the same gold substrate, a matrix-free C1 chip, and a CM5 chip with a ~100 nm dextran hydrogel matrix (GE Healthcare). Two well-studied interaction types were tested, the binding of a biotinylated antibody (immunoglobulin G) to streptavidin and an antigen-antibody interaction. For both interactions, the well characterized densely functionalized SAM-MHDA chip gave a high signal-to-noise ratio and showed a gain in the availability of immobilized ligands for their partners injected in buffer flow, thus showing efficacy over commercially available sensor chips [96].

14.21. Microfabricatable Mid-infrared (mid-IR) Surface Plasmon Resonance (SPR)

The proposed platform has a flat grating SPR coupler and periodic heavily doped profiles implanted into intrinsic silicon and a thin gold layer deposited on top. Results show that the proposed structure can excite the SPR on the normal incidence of mid-IR light, resulting in a large probing depth that will facilitate the study of larger analytes. Whole structure can be micro fabricated with other batch protocols, providing tunability in the SPR excitation wavelength for specific biosensing needs with a low manufacturing cost [97].

14.22. Fiber Optic Particle Plasmon Resonance (FOPPR) Biosensor

The FOPPR sensor is based on gold nano-particle modified optical fiber. The bimolecular binding event was monitored in real-time. Assuming a Langmuir-type adsorption isotherm, initial binding rate, quantitative determination of binding kinetic constants, the association and dissociation rate constants were determined thus showing its efficacy for studying real-time biomolecular interactions [98].

14.23. (2D) Spectral SPR

Two-dimensional (2D) spectral SPR sensor is based on a polarization control scheme. The polarization control configuration converts the phase difference between p and s polarization occurring at SPR into corresponding color responses in spectral SPR images with a sensor resolution of 2.7×10^{-6} RIU. The improvement in magnitude resolution is about 26 times compared to the existing 2D spectral SPR sensors [99].

14.24. Nanoplasmonic Microfluidic (NMF) Transmission Surface Plasmon Resonance (SPR) Biosensor

Malic et al. developed an all-polymeric NMF transmission SPR biosensor. The device was fabricated in thermoplastics by hot embossing technique. The fabrication process enabled monolithic integration of blazed nano-gratings within the detection chambers of a multichannel microfluidic system. As a result, single hard thermoplastic bottom substrate comprising plasmonic and fluidic features allowed integration of active fluidic elements, such as pneumatic valves, in the top soft thermoplastic cover, increasing device functionality. A simple and compact transmission-based optical setup was employed with multiplexed end-point or dual-channel kinetic detection capability which did not require stringent angular accuracy [100].

14.25. Wavelength Interrogation Surface Plasmon Resonance (SPR) Spectroscopy

Wavelength interrogation SPR spectroscopy uses a dove prism and integrates optical design with high-resolution refractive index monitoring and biosensing. This optical configuration uses a single axis optical path between each optical component with incorporation of fluidics for efficient sample delivery [87].

14.26. Surface Plasmon Resonance-phase Imaging (SPR-PI) and Nanoparticle-Enhanced SPR-PI

SPR-PI experiments utilized a near infrared 860 nm LED light source and a wedge depolarizer to create a phase grating on a four-element single-stranded DNA (ssDNA) microarray. Multiplexed real time phase shift measurements were used to detect bio-affinity adsorption onto the various microarrays. Adsorbed silica nano-particles provided a greatly enhanced phase shift upon bio-affinity adsorption due to a large increase in the real component of the interfacial refractive index from the adsorbed nano-particle. Detection was 20 times more sensitive than that observed previously with nano-particle-enhanced SPRI [101].

14.27. Other Applications of SPR Biosensors

A SPR for high-speed microfluidic delivery of liquid samples to sensor surface was designed. This consist of four-port microfluidic cell, two ports serve as inlets for buffer and sample solutions respectively while a high-speed selector valve to control the alternate opening and closing of the two outlet ports. The time scale of buffer/sample switching (or sample injection rise and fall time) is on the order of milliseconds. This minimizes the opportunity for sample plug dispersion. The high rates of mass transport allowed for kinetic analysis of binding events with dissociation rate constants (kd) up to 130 $s^{-1}$with the sample requirement of 1μL, thus allowing for minimal sample consumption during high-speed kinetic binding measurement. [102].

A novel SPR biosensing platform incorporating temporal carrier technique with common-path spectral interferometry was designed. This platform achieves the refractive index resolution (RIR) up to 2×10^{-8} RIU over a wide dynamic range of 3×10^{-2} RIU. This is accomplished by optimizing the SPR differential phase sensing conditions with a gold layer to address the spectral phase discontinuity with a novel spectral-temporal phase measurement scheme. The proposed optical setup supersedes its Michelson counterpart in term of simplicity and presumed to have a significant contribution for practical SPR sensing applications [103].

A novel *planar plasmonic interferometers* based biochemical sensor consisting of nano-scale grooves and slits in a metal film to form two-arm, three-beam was designed. Thousands of plasmonic interferometers per square millimeter were integrated with a microfluidic system to detect physiological concentrations of glucose in water over a broad wavelength range of 400-800 nm. Wavelength sensitivity between 370 and 630 nm / RIU, a relative intensity change between $\sim 10^3$ and 10^6 % / RIU, and a resolution of $\sim 3 \times 10^{-7}$ in refractive index change were measured using sensing volumes as low as 20 fL. This clearly demonstrated that multispectral plasmonic interferometry is a promising approach for the development of high-throughput, real-time, and extremely compact biochemical sensors [104].

Using the combine of three different transduction techniques, viz. Polarization modulation-reflection absorption infrared spectroscopy (PM-RAIRS), quartz crystal microbalance with dissipation measurements (QCM-D), and Fourier transform-surface plasmon resonance (FT-SPR), immobilization of anti-rabbit immunoglobulins (anti-rIgGs in concentrations of 1-50 μg/ml) and the detection of rIgGs were monitored on gold transducers. All three techniques could detect antibody binding in the low concentration of 1 μg/ml. Of these, FT-SPR leads to an intense signal with a wavenumber shift of approximately 30 cm-1 thus expecting a detection limit of the order of a few tenths of μg /ml by FT-SPR. [105].

CONCLUSION

SPR provides a indispensible and powerful tool for the analysis of protein/protein interactions even for low affinity interactions, which are difficult to study using any other technique. One of the most distinctive features of SPR is in providing information on structural integrity of recombinant molecules through binding analysis. SPR also measures the affinity, enthalpy, stoichiometry, kinetics and activation energy of an interaction. SPR require small amount of protein as compared to other techniques such as calorimetry. The drawbacks associated with SPR are easily avoided once they are understood. However, SPR is not well-suited to high-throughput assays, or the analysis of small molecules ($M_r < 1000$).

Chapter 15

APPENDIX [23]

15.1. PHYSICAL BASIS OF SPR

When a beam of light passes from material with a high refractive index (e.g. glass) into material with a low refractive index (e.g. water) some light is reflected from the interface. When the angle at which the light strikes the interface (the angle of incidence or θ) is greater than the critical angle (θ_c), the light is completely reflected (total internal reflection). If the surface of the glass is coated with a thin film of a noble metal (e.g. gold), this reflection is not total; some of the light is 'lost' into the metallic film. There then exists a second angle greater than the critical angle at which this loss is greatest and at which the intensity of reflected light reaches a minimum or 'dip'. This angle is called the surface plasmon resonance angle (θ_{spr}). It is a consequence of the oscillation of mobile electrons (or 'plasma') at the surface of the metal film. These oscillating plasma waves are called surface plasmons. When the wave vector of the incident light matches the wavelength of the surface plasmons, the electrons 'resonate', hence the term surface plasmon resonance. The 'coupling' of the incident light to the surface plasmons results in a loss of energy and therefore a reduction in the intensity of the reflected light. It is because the amplitude of the wave vector in the plane of the metallic film depends on the angle at which it strikes the interface that, θ_{spr} is observed. An evanescent (decaying) electrical field associated with the plasma wave travels for a short distance (~300 nm) into the medium from the metallic film. Because of this, the resonant frequency of the surface plasma wave (and thus θ_{spr}) depends on the refractive index of this medium. If the surface is immersed in an aqueous buffer (refractive index or μ ~1.0) and protein (μ ~1.33) binds to the surface, this results in an increase in refractive index which is detected by a shift in the θ_{spr}. The instrument uses a photo-detector array to measure very small changes in θ_{spr}. The readout from this array can be viewed on the SPR instruments as 'dips' (Section 6.1.3). The change is quantified in resonance units or response units (RUs) with 1 RU equivalent to a shift of 10^{-4} degrees. Empirical measurements have shown that the binding of 1 ng/mm^2 of protein to the sensor surface leads to a response of ~1000 RU. Since the matrix is ~100 nm thick, this represents a protein concentration within the matrix of 10 mg/mL. Apart from the refractive index, the other physical parameter which affects θ_{spr} is temperature. Thus, a crucial feature of any SPR instrument is precise temperature control.

15.2. Samples and Buffers [23]

15.2.1. Guidelines for Preparing Samples and Buffers for SPR Experiments

- All buffers should be filtered through 0.2 μm filters and degassed at room temperature. The latter can be achieved by filtering under vacuum or by using a vacuum chamber.
- All samples with volume greater than 3 mL should be filtered and degassed in the same way.
- If sample volume is less than 3 mL, it should be spun at high speed in a micro centrifuge for 10-20 min at 4°C and then degassed in a vacuum chamber.
- Before samples are placed in the sample rack they should be pulsed briefly in a micro centrifuge. This dislodges air-bubbles from the bottom of the container; helps ensure that the meniscus is horizontal.
- Vials should be capped to prevent sample evaporation.
- When running long experiments consider cooling the sample rack base using a thermostatic re-circulator.

15.2.2. Standard Buffers [23]

Running Buffers

HBS or HBS-EP: 10 mM HEPES pH 7.4, 150 mM NaCl, 3 mM EDTA, 0.005% P20

HBS-P: 10 mM HEPES pH 7.4, 150 mM NaCl, 0.005% P20

HBS-N: 10 mM HEPES pH 7.4, 150 mM NaCl

Pre-concentration Buffers

pH 3.0-4.5 10 mM formate

pH 4.0-5.5 10 mM acetate

pH 5.5-6.0 5 mM maleate

Regeneration Buffers

Grouped according to chemical properties. They increase in strength from left to right.

Cation chelator: HBS with 20 mM EDTA pH 7.5

High ionic strength: NaCl 1 M; KCl 4 M; $MgCl_2$ 2M

Low pH: Glycine/HCl 100 mM pH 2.5; HCl 10 mM; HCl 100 mM; H_3PO_4 100 mM

High pH: NaOH 5 mM; NaOH 50 mM

References

[1] Willets, K.A. & Van Duyne, R.P. (2007). Localized surface plasmon resonance spectroscopy and sensing. *Annu Rev Phys Chem. 58*, 267-97.

[2] Drude, P. (1900). Zur Elektronentheorie der Metalle; II. Teil. Galvanomagnetische und thermomagnetische Effecte. *Annalen der Physik 308*, 369.

[3] Rich, R.L. & Myszka, D.G. (2007). Higher throughput, label-free, real-time molecular interaction analysis. *Anal biochem 361*, 1–6.

[4] Stefan, A., Meyer, Baptiste Auguié, Eric, C., Le R. & Pablo, G. (2012). Etchegoin Combined SPR and SERS Microscopy in the Kretschmann Configuration. *J Phys Chem A* 116, 1000–1007.

[5] Otto, A. (1968). A new method for exciting non-radioactive surface plasma oscillations. *Phys Stat Sol 26*, 99-101.

[6] Otto, A. (1968). Excitation of nonradioactive surface plasma waves in silver by the method of frustrated total reflection. *Z Phys 216*, 398-410.

[7] Zeng, S., Yu, X., Law, W.C., Zhang. Y. Hu R., Dinh, X.Q., Ho, H.P. & Yong, K.T. (2013). Size dependence of Au NP-enhanced surface plasmon resonance based on differential phase measurement. *Sens Actu B: Chem 176,*1128-1133.

[8] Stefan, M. *Plasmonics: Fundamentals and Applications*. Springer; (2007).

[9] Richard, B.M.S. Handbook of Surface Plasmon Resonance. In: Tudos, AJ (Editor) RSC publishing, 2008.

[10] Jönsson, U., Fägerstam, L., Ivarsson, B., Johnsson, B., Karlsson, R., Lundh, K., Löfås, S., Persson, B., Roos, H., Rönnberg, I., Sjölander, S., Stenberg, E., Ståhlberg, R., Urbaniczky, C., Östlin, H. & Malmqvist, M. (1991). Real-time biospecific interaction analysis using surface plasmon resonance and a sensor chip technology. *BioTech 11*, 620-627.

[11] Jönsson, U. & Malmqvist, M. (1992). Advances in Biosensors In: Turner, A (Editor). San Diego: JAI Press; 1992; 291-336.

[12] Shin, W., Matsumiya, M., Qiu, F., Izu, N. & Murayama, N., (2004). Thermoelectric gas sensor for detection of high hydrogen concentration. *Sens Actu B 97*, 344–347.

[13] Takeuchi, K., Tanaka, T., Ikeda, M., Shibata, K., Sakauchi, Y., Yamada, Y. & Nakano, S. (1993). Highly accurate CO_2 gas sensor using a modulation-type pyroelectric infrared detector. *Jpn J Appl Phys 32*, 221–227.

[14] Sigrist, M.W. (1995). Trace gas monitoring by laser-photoacoustic spectroscopy. *Inf Phys Technol 36*, 415–425.

[15] Hodgson, J. (1994). Light, Angles, Action. Instruments for label free, real-time monitoring of intermolecular interactions. *Biotech 12*, 31-35.

[16] van der Merwe, P.A. & Barclay, A.N. (1996). Analysis of cell-adhesion molecule interactions using surface plasmon resonance. *Curr Opin Immunol 8*, 257-261.

[17] Schuck, P. (1997). Use of surface plasmon resonance to probe the equilibrium and dynamic aspects of interactions between biological macromolecules. *Annu Rev Biophys Biomol Struct 26*, 541-566.

[18] O'Shannessy, D.J. (1994). Determination of kinetic rates and equilibrium binding constants for macromolecular interactions. A critique of Surface Plasmon Resonance Literature. *Curr Opin Biotechnol 5*, 65-71.

[19] van der Merwe, P. A., Brown, M. H., Davis, S. J. & Barclay, A. N. (1993). Measuring very low affinity interactions between immunoglobulin superfamily cell-adhesion molecules. *Biochem Soc Trans 21*, 340.

[20] Myszka, D.G. (1997). Kinetic analysis of macromolecular interaction using surface Plasmon resonance biosensors. *Curr Opin Biotechnol 8*, 50.

[21] Karlsson, R. & Falt, A. (1997). Experimental design for kinetic analysis of protein-protein interactions with surface plasmon resonance biosensors. *J Immunol Meth 200*, 121-133.

[22] van der Merwe, P. A. & Barclay, A.N. (1994). Transient intercellular adhesion: the importance of weak protein-protein interactions. *Tr Biochem Sci 19*, 354-358.

[23] P. Anton van der Merwe. SPR. http://www.academia.edu/ 4406483/ Surface Plasmon resonance

[24] Herminjard, S., Sirigu, L., Herzig, H.P., Studemann, E., Crottini, A., Pellaux, J.P., Gresch, T., Fische, M. & Faist, J. (2009). Surface Plasmon Resonance sensor showing enhanced sensitivity for CO_2 detection in the mid-infrared range. *Opt Expr 17*: 293-303.

[25] Anderson, K., Hamalainen, M. & Malmqvist, M. (1999). Identification and optimization of regeneration conditions for affinity-based biosensor assays. A multivariate cocktail approach. *Anal Chem 71*, 2475-2481.

[26] Maren, T.H. & Conroy, C.W. (1993). A new class of carbonic anhydrase inhibitor. J Biol Chem 268, 26233-26239.

[27] Papalia, G.A., Leavitt, S., et al. (2006). Comparative analysis of 10 small molecules binding to carbonic anhydrase II by diVerent investigators using Biacore technology. *Anal Biochem 359*, 94-105.

[28] Guiseppi-Ellie, A., Pradham, S.R. & Wilson, A.M. (1993). Growth of electropolymerized polyaniline thin films. *Chem Mater 5*: 1474-1480.

[29] Baba, A., Tian, S. et al. (2004). Electropolymerization and doping/dedoping properties of polyaniline thin films as studied by electrochemical-surface plasmon spectroscopy and by the quartz crystal microbalance. *J Electroanal Chem 562*, 95-103.

[30] Duval, A., Aide, A., Bellemain, A., Moreau, J., Canva, M. & Bardin, F. (2007). Anisotropic surface-plasmon- resonance imaging biosensor. *SPIE* 2-3, doi: 10.1117/2.1200710.0882.

[31] Heaton, R.J., Peterson, A.W. & Georgiadis, R.M. (2001). Electrostatic surface plasmon resonance: Direct electric field-induced hybridization and denaturation in monolayer nucleic acid films and label-free discrimination of base mismatches. *Proc Natl Acad Sci 98, 3701–3704.*

[32] Miura, S., Watanabe, O.K., Nishizawa, S. & Teramae, N. (2008). A surface Plasmon resonance sensor based on 3-5 diaminopyrazine with a high selectivity for thymine in AP-site containing nucleobases in DNA duplexes. *Nuc Ac Symp Ser 52*, 123-124.

[33] Hutter, E. & Fendler, J.H. (2004). Exploitation of Localized Surface Plasmon Resonance. *Adv Mater 16*, 1685-1706.

[34] Wyer, J.R., Willcox, B.E., Gao, G.F., Gerth, U.C., Davis, S.J., Bell, J. I., van der Merwe, P. A. & Jakobsen, B.K. (1999). T cell receptor and co-receptor CD8 alpha bind peptide-MHC independently and with distinct kinetics. *Immunity 10*, 219-225.

[35] Willcox, B.E., Gao, G.F., Wyer, J.R., Ladbury, J.E., Bell, J.I., Jakobsen, B.K. & van der Merwe, P.A. (1999). TCR binding to peptide-MHC stabilizes a flexible recognition interface. *Immunity 10*, 357-365.

[36] van der Merwe, P.A., Crocker, P., Vincent, M., Barclay, A.N. & Kelm, S. (1996). Localization of the Putative Sialic Acid-binding Site on the Immunoglobulin Superfamily Cell-surface Molecule CD22. J Bio Chem 271, 9273-9280.

[37] van der Merwe, P.A., McNamee, P.N., Davies, E.A., Barclay, A.N. & Davis, S.J. (1995). Topology of the CD2-CD48 cell-adhesion molecule complex: implications for antigen recognition by T cells. *Curr Biol 5*, 74.

[38] Davis, S. J., Davies, E. A., Tucknott, M. G., Jones, E. Y., & van der Merwe, P. A. (1998). The role of charged residues mediating low affinity protein-protein recognition at the cell surface by CD2. *Proc Natl Acad Sci USA 95*, 5490-5494.

[39] Woan, G. The Cambridge Handbook of Physics Formulas. Cambridge University Press, 2010.

[40] Griffiths, D.J. Introduction to Electrodynamics. Pearson Education; 2007.

[41] Gustav, M. (1908). Beiträge zur Optik trüber Medien, speziell kolloidaler Metallösungen. *Annalen der Physik 330*, 377–445.

[42] Stratton, J.A. Electromagnetic Theory. New York: McGraw-Hill; 1941.

[43] van der Merwe, P.A., Bodian, D. L., Daenke, S., Linsley, P. & Davis, S.J. (1997). CD80 (B7-1) binds both CD28 and CTLA-4 with a low affinity and very fast kinetics. *J Exp Med 185*, 393-403.

[44] van der Merwe, P.A., Brown, M.H., Davis, S.J., Barclay & A.N. (1993). Affinity and kinetic analysis of the interaction of the cell adhesion molecules rat CD2 and CD48. *EMBO J 12*, 4945-4954.

[45] van der Merwe, P.A., Barclay, A.N., Mason, D.W., Davies, E.A., Morgan, B.P., Tone, M., Krishnam, A.K.C., Ianelli, C. & Davis. S.J. (1994). Human cell-adhesion molecule CD2 binds CD58 (LFA-3) with a very low affinity and an extremely fast dissociation rate but does not bind CD48 or CD59. *Biochem 33*, 10149-10160.

[46] Karlsson, R. & Ståhlberg, R. (1995). Surface plasmon resonance detection and multispot sensing for direct monitoring of interactions involving low-molecular-weight analytes and for determination of low affinities. *Anal Biochem 228*, 274-280.

[47] Nicholson, M.W., Barclay, A.N., Singer, M.S., Rosen, S.D. & van der Merwe, P.A. (1998). Affinity and kinetic analysis of L-selectin (CD62L) binding to glycosylation-dependent cell-adhesion molecule-1. *J Biol Chem 273*, 763-770.

[48] George, W. D. (1984). Constructing nonlinear Scatchard plots. *J Chem Educ* 61, 527.

[49] Gravell, M., London, W.T., Houf, S.A., Madden, D.L., Dalakas, M.C. & Sever, J.L. (1984). Analyzing nonlinear scatchard plots. *Science 223*, 76-78.

[50] 21. Felder, S., Zhou, M., Hu, P., Ureña, J., Ullrich, A., Chaudhuri, M., White, M., Shoelson, S.E. & Schlessinger, J. (1993). SH2 domains exhibit high-affinity binding to tyrosine-phosphorylated peptides yet also exhibit rapid dissociation and exchange. *Mol Cell Biol 13*, 1449-1455.

[51] Myszka, D.G. & Morton, T.A. (1998). CLAMP: a biosensor kinetic data analysis program. *Tr Biochem Sc 23*,149-150.

[52] Yoo, S.H. & Lewis, M.S. (1995). Thermodynamic study of the pH-dependent interaction of chromogranin A with an intraluminal loop peptide of the inositol 1,4,5-trisphosphate receptor. *Biochem 34*, 632-680.

[53] Stites, W.E. (1997). Proteinminus signProtein Interactions: Interface Structure, Binding Thermodynamics, and Mutational Analysis. *Chem Rev 97*, 1233-1250.

[54] Ladbury, J.E. & Chowdhry, B.Z. (1996). Sensing the heat: the application of isothermal titration calorimetry to thermodynamic studies of biomolecular interactions. *Chem Biol 3*, 791-801.

[55] Naghibi, H., Tamura, A. & Sturtevant, J.M. (1995). Significant discrepancies between van't Hoff and calorimetric enthalpies. *Proc Natl Acad Sci U S A 92*, 5597-5599.

[56] Liedeberg, B., Nylander, C. & Lundström I. (1995). Biosensors Bioelectron. *10*, 1–9.

[57] Alves, I.D., Park, C.K. & Hruby, V.J. (2005). Plasmon Resonance Methods in GPCR Signaling and Other Membrane Events. *Curr Protein Pept Sci. 6*, 293–312.

[58] Uetz P., Loic, G., et al. (2000). A comprehensive analysis of protein–protein interactions in *Saccharomyces cerevisiae. Nature* 403, 623–627.

[59] Hyungsoon, I.M., Sutherland, J.N., Maynard, J.A. and Sang-Hyun, O.H. (2012). Nanohole-based SPR Instruments with Improved Spectral Resolution Quantify a Broad Range of Antibody-Ligand Binding Kinetics. *Anal Chem. 84*, 1941–1947.

[60] Menezes, J.W., Ferreira, J., Santos, M.J.L., Cescato, L. & Brolo, A.G. (2010). *Adv. Funct. Mater. 20*, 3918–3924.

[61] Dahlin, A., Zäch, M., Rindzevicius, T., Käll, M., Sutherland, D.S. & Höök, F. J. (2005). *Am Chem Soc* 127, 5043–5048.

[62] Hyungsoon, I.M., Lesuffleur, A., Lindquist, N.C. & Sang-Hyun, O.H. (2009). Plasmonic Nanoholes in a Multi-Channel Microarray Format for Parallel Kinetic Assays and Differential Sensing. *Anal Chem. 81*, 2854–2859. doi: 10.1021/ac802276x

[63] Jonsson, M.P., Dahlin, A.B., Feuz, L., Petronis, S. & Höök, F. (2010). *Anal. Chem. 82*, 2087–2094.

[64] Hyungsoon, I.M., Sutherland, J.M., Maynard, J.A. & Sang-Hyun, O.H. (2012). Nanohole-based SPR Instruments with Improved Spectral Resolution Quantify a Broad Range of Antibody-Ligand Binding Kinetics, *Anal Chem. 84*, 1941–1947. Published online 2012 February 7. doi: 10.1021/ac300070t

[65] Lee, K.L., Wu, S.H., Lee, C.W. & Wei, P.K. (2011). Sensitive biosensors using Fano resonance in single gold nanoslit with periodic grooves. *Opt Expr* 19, 24530-9. doi: 10.1364/OE.19.024530.

[66] Neumann, T., Junker, H.D., Schmidt, K. & Sekul, R. (2007). SPR-based fragment screening: advantages and applications. *Curr Top Med Chem* 7, 1630-42.

[67] Maynard, J.A., Lindquist, N.C., Sutherland, J.N., Lesuffleur, A., Warrington, A.E., Rodriguez, M. & Sang-Hyun O.H. (2009). Next generation SPR technology of membrane-bound proteins for ligand screening and biomarker discovery. *Biotechnol J 4*, 1542–1558. doi: 10.1002/biot.200900195

[68] Lee, S.H., Lindquist, N.C., Wittenberg, N.J., Jordan, L.R. & Oh, S.H. (2012). Real-time full-spectral imaging and affinity measurements from 50 microfluidic channels using nanohole surface plasmon resonance. *Lab Chip 12*, 3882-3890.

[69] Wittenberg, N.J., Im, H., Johnson, T.W., Xu, X., Warrington, A.E., Rodriguez, M., Oh, S.H. (2011). *ACS Nano 5*, 7555–7564.

[70] Lindquist, N.C., Nagpal, P., McPeak, K.M., Norris, D.J. & Oh, S.H. (2012). *Rep Prog Phys* 75036501

[71] Yu, C.C., Ho, K.H., Chen, H.L., Chuang, S.Y., Tseng, S.C. & Su, W.F. (2012). Using the nanoimprint-in-metal method to prepare corrugated metal structures for plasmonic biosensors through both surface plasmon resonance and index-matching effects. *Biosens Bioelectron 33*, 267-73.

[72] Xue, J., Zhou, W., Dong, B., Wang, X., Chen, Y., Huq, E., Zeng, W., Qu, X. & Liu, R. (2009). Surface plasmon enhanced transmission

through planar gold quasicrystals fabricated by focused ion beam technique. *Microelectron Eng 86*, 1131–1133.

[73] Sharpe, J.C., Mitchell, J.S., Lin, L., Sedoglavich, N. & Blaikie, R.J. (2008). Gold nanohole array substrates as immune-biosensors. *Anal Chem 80*, 2244–2249.

[74] Blanchard-Dionne, A.P., Guyot, L., Patskovsky, S., Gordon, R. & Meunier, M. (2011). Intensity based surface plasmon resonance sensor using a nanohole rectangular array. *Opt Expr 19*, 15041-6.

[75] Lazzara, T.D., Carnarius, C., Kocun, M., Janshoff, A. & Steinem, C. (2011). *ACS Nano 5*, 6935–6944.

[76] Nakamoto, K., Kurita, R., Niwa, O., Fujii, T. & Nishida, M. (2011). Development of a mass-producible on-chip plasmonic nanohole array biosensor. *Nanoscale* 3, 5067-75.

[77] Couture, M., Liang, Y., Poirier Richard, H.P., Faid, R., Peng, W. & Masson, J.F. (2013). Tuning the 3D plasmon field of nanohole arrays. *Nanoscale 5*, 12399-408.

[78] Martinez-Perdiguero, J., Retolaza, A., Otaduy, D., Juarros, A. & Merino, S. (2013). Real-Time Label-Free Surface Plasmon Resonance Biosensing with Gold Nanohole Arrays Fabricated by Nanoimprint Lithography. *Sensors 13*, 13960-13968.

[79] Nedelkov, D., Tubbs, K.A. & Nelson, R.W. (2006). Surface plasmon resonance-enabled mass spectrometry arrays. *Electrophoresis 27*, 3671-5.

[80] Bellon, S., Buchmann, W., Gonnet, F., Jarroux, N., Anger-Leroy, M., Guillonneau, F. & Daniel, R. (2009). Hyphenation of surface plasmon resonance imaging to matrix-assisted laser desorption ionization mass spectrometry by on-chip mass spectrometry and tandem mass spectrometry analysis. *Anal Chem 81*, 7695-702.

[81] Dhawan, A., Canva, M. & Vo-Dinh, T. (2011). Narrow groove plasmonic nano-gratings for surface plasmon resonance sensing. *Opt Expr 19*, 787–813.

[82] Campbell, C.T. & Kim, G. (2007). SPR microscopy and its applications to high-throughput analyses of biomolecular binding events and their kinetics. *Biomaterials 28*, 2380-92.

[83] Wark, A.W., H.J. Lee, & Corn, R.M. Advanced Methods for SPR Biosensing. In: R.B.M. Schasfoort, R.B.M. & Tudos, A.J. (Eds.), Handbook of Surface Plasmon Resonance. Cambridge : RSC Publishing; 2008; 251-280.

[84] Stojanović, I., Schasfoort, R.B. & Terstappen, L.W. (2014). Analysis of cell surface antigens by Surface Plasmon Resonance imaging. *Biosens Bioelectron* 52, 36-43.

[85] Stiles, P.L., Dieringer, J.A., Shah, N.C. & Van Duyne, R.P. (2008). Surface-enhanced Raman spectroscopy. *Annu Rev Anal Chem (Palo Alto Calif)* 1, 601-26.

[86] Williams, L.D., Ghosh, T., Mastrangelo, C.H. (2010). Low noise detection of biomolecular interactions with signal-locking surface plasmon resonance. *Anal Chem* 82, 6025-31.

[87] Bolduc, O.R., Live, L.S. & Masson, J.F. (2009). High-resolution surface plasmon resonance sensors based on a dove prism. *Talanta 77*, 1680-7.

[88] Ng, S.P., Wu, C.M., Wu, S.Y., Ho, H.P. & Kong, S.K. (2010). Differential spectral phase interferometry for wide dynamic range surface plasmon resonance biosensing. *Biosens Bioelectron 26*, 1593-8.

[89] Patskovsky, S., Latendresse, V., Dallaire, A.M., Doré-Mathieu, L., Meunier, M. (2014). Combined surface plasmon resonance and impedance spectroscopy systems for biosensing. *Analyst 139,* 596-602.

[90] Subramanian, K., Fee, C.J., Fredericks, R., Stubbs, R.S. & Hayes, M.T. (2013). Insulin receptor-insulin interaction kinetics using multiplex surface plasmon resonance. *J Mol Recognit 26*, 643-52.

[91] Liu, X., Du, M., Zhou, F., Gomez, F.A. (2012). Facile fabrication of an interface for online coupling of microchip CE to surface plasmon resonance. *Bioanalysis 4*, 373-9.

[92] Tassa, C., Liong, M., Hilderbrand, S., Sandler, J.E., Reiner, T., Keliher, E.J., Weissleder, R. Shaw, S.Y. (2013). Microfluidic on-chip capture-cycloaddition reaction to reversibly immobilize small molecules or multi-component structures for biosensor applications. *J Vis Exp 79*, :e50772. doi: 10.3791/50772.

[93] de Boer, A.R., Hokke, C.H., Deelder, A.M. & Wuhrer, M. (2008). Serum antibody screening by surface plasmon resonance using a natural glycan microarray. *Glycoconj J 25*, 75-84.

[94] Glück, J.M., Koenig, B.W. & Willbold, D. (2011). Nanodiscs allow the use of integral membrane proteins as analytes in surface plasmon resonance studies. *Anal Biochem 408*, 46-52.

[95] Kim, J.A., Kulkarni, A., Kang, J., Amin, R., Choi, J.B., Park, S.H. & Kim, T. (2012). Evaluation of multi-layered graphene surface plasmon resonance-based transmission type fiber optic sensor. *J Nanosci Nanotechnol 12*, 5381-5.

[96] Roussille, L., Brotons, G., Ballut, L., Louarn, G., Ausserré, D. & Ricard-Blum, S. (2011). Surface characterization and efficiency of a matrix-free and flat carboxylated gold sensor chip for surface plasmon resonance (SPR). *Anal Bioanal Chem 401*, 1601-17.

[97] DiPippo, W., Lee, B.J. & Park, K. (2010). Design analysis of doped-silicon surface plasmon resonance immunosensors in mid-infrared range. *Opt Expr 18*, 19396-406.

[98] Chang, T.C., Wu, C.C., Wang, S.C., Chau, L.K., Hsieh & W.H. (2013). Using a fiber optic particle plasmon resonance biosensor to determine kinetic constants of antigen-antibody binding reaction. *Anal Chem 85*, 245-50.

[99] Wong, C.L., Chen, G.C., Ng, B.K., Agarwal, S., Lin, Z., Chen, P. & Ho, H.P. (2011). Multiplex spectral surface plasmon resonance imaging (SPRI) sensor based on the polarization control scheme. *Opt Expr 19*, 18965-78.

[100] Malic, L., Morton, K., Clime, L. & Veres, T. (2013). All-thermoplastic nanoplasmonic microfluidic device for transmission SPR biosensing. *Lab Chip 13*, 798-810.

[101] Zhou, W.J., Halpern, A.R., Seefeld, T.H. & Corn, R.M. (2012). Near infrared surface plasmon resonance phase imaging and nanoparticle-enhanced surface plasmon resonance phase imaging for ultrasensitive protein and DNA biosensing with oligonucleotide and aptamer microarrays. *Anal Chem 84*, 440-5.

[102] Macgriff, C.A., Wang, S. & Tao, N. (2013). Four-port microfluidic flow-cell with instant sample switching. *Rev Sci Instrum 84*,106-110.

[103] Ng, S.P., Loo, F.C., Wu, S.Y., Kong, S.K., Wu, C.M. & Ho, H.P. (2013). Common-path spectral interferometry with temporal carrier for highly sensitive surface plasmon resonance sensing. *Opt Expr 21*, 20268-73.

[104] Feng, J., Siu, V.S., Roelke, A., Mehta, V., Rhieu, S.Y., Palmore, G.T. & Pacifici, D. (2012). Nanoscale plasmonic interferometers for multispectral, high-throughput biochemical sensing. *Nano Lett 12*, 602-9.

[105] Boujday, S., Méthivier, C., Beccard, B. & Pradier, C.M. (2009). Innovative surface characterization techniques applied to immunosensor elaboration and test: comparing the efficiency of Fourier transform-surface plasmon resonance, quartz crystal microbalance with dissipation measurements, and polarization modulation-reflection absorption infrared spectroscopy. *Anal Biochem 387*, 194-201.

INDEX

#

A

B

H

I

J

K

L

M

N

P

Q

R

S

T

U

V

W

X

Y

Δ